# Charles S. Peirce's
## Mathematical Logic
## and Philosophy

Alan James Iliff

Docent
Press

DOCENT PRESS
Boston, Massachusetts, USA
www.docentpress.com

Docent Press publishes books in the history of mathematics, computing, and technology about interesting people and intriguing ideas. The histories are told at many levels of detail and depth that can be explored at leisure by the general reader.

Cover design by Brenda Riddell, Graphic Details.

Produced with TeX.

9 8 7 6 5 4 3 2 1

ISBN-10: 1-942795-99-8
ISBN-13: 978-1-94275-99-5

# Contents

# Acknowledgments

The work on adapting my doctoral thesis to become this book was partially supported by a North Park University sabbatical granted during the fall semester of 2017.

I am grateful to Dennis Cowan, then Dean of Faculty at Shimer College, for introducing me to the writings of Charles S. Peirce in the fall of 1969.

I acknowledge with gratitude the help of all the logicians who have taught and counselled me. Most important of all were the valuable time and efforts of my adviser, William A. Howard.

Without these individuals and many others, I might have become even more literary and metaphysical, even less scientific and mathematical.

> *Poi ch'innalzai un poco più le ciglia,*
> *vidi 'l maestro di color che sanno*
> *seder tra filosofica famiglia.*

> When I raised my eyes a little higher,
> I saw the Master of those who know,
> seated in a philosophic family.

*Inferno IV* (Dante, 1980, pp. 42–43)

*This book is dedicated to my parents, James and Veta, and to my wife, Lillian.*

# New Preface (2018)

My original thesis was completed and submitted to the University of Illinois at Chicago in December of 1992. When Docent Press offered to publish my doctoral thesis, I was quite pleasantly surprised. The possibility of using a Sabbatical Leave from North Park University to do so was a blessing for which I am very grateful, and working with Docent Press has been highly gratifying. I decided not to expand or to revise the original document, both because it has entered into the extant literature on Peirce,[1] and because no new developments have altered the evidence upon which it is based. Therefore, I limited myself to making many small improvements and clarifications, which included streamlining some of the titles of chapters and sections.

The great International Congress for Charles S. Peirce studies in 1989 at Harvard was a peak experience for me, since it came in the midst of writing my doctoral thesis on Peirce's mathematical logic. At that Congress I was able to read a paper based on some partial results of my work, which was subsequently published in a volume of papers on similar topics.[2]

It seems that the Congress was, in a certain sense, the proximate cause of bringing to light the excellent biography by Joseph Brent, which was his 1960 dissertation in History at UCLA, and which was finally published in 1993.[3] It was eagerly read by Peirce scholars, and I recommend it enthusiastically to anyone who is interested in Peirce. (The story of how Brent was found after many years is told by Thomas Sebeok in the front matter.) After some long suppressed materials were finally pried loose from those who had withheld the materials for decades, Brent was able to

---

[1] In fact, after running across the term "Alan Iliff's Conjecture," I was somewhat bemused to discover with a search engine that it refers to my speculations in Section 6.2 about Virgil's *Aeneid* and the multiple meanings that can be ascribed to the name "Arisbe", which Peirce gave to his home in Pennsylvania. Unless and until contemporary testimony on this from Peirce or any of his close companions is found, it must remain merely conjectural.

[2] Iliff, A.: The Role of the Matrix Representation in Peirce's Development of the Quantifiers. In *Studies in the Logic of Charles Sanders Peirce*. Ed. Houser, N., Roberts, D., and Van Evra, J. Bloomington, Indiana University Press, 1997.

[3] Brent, J.: *Charles Sanders Peirce: A Life*. Bloomington, Indiana University Press, 1993

publish an even more perfect second edition,[4] and Brent's account of his remarkable experiences trying to tell the true story about Peirce's difficult and unfortunate career was published online.[5]

Some time after I returned home from the Peirce Congress, my thesis adviser, Bill Howard, said words to the effect that the story of how Peirce's work was disparaged and belittled is so complicated that it would take a whole dialogue to lay it all out thoroughly. At the time, I thought he was merely joking, but to my surprise I discovered that he was serious, when the next time we met he asked me what ideas I had come up with for the dialogue! We both felt that the resulting Chapter 6 was quite successful, both in form and substance.

The researches of Irving Anellis have been most valuable in making known the extent of Peirce's work, and I especially recommend his online article, 'How Peircean was the "Fregean Revolution" in Logic?'[6]

The importance of the two lines of development, deduction-theoretic and model-theoretic is central to the task of finding Peirce's place in the history of mathematical logic, and the name of Jaakko Hintikka should well be added to the names of those who have identified the model-theoretic line of development, as discussed at the end of Chapter 6.[7]

It was in the online paper by Anellis that I found a reference to the following passage from the beginning of a paper by Alfred Tarski[8], and I would certainly have included this quotation at the end of Section 6.2 if I had known of its existence. Tarski is one of the creators, if not the chief creator, of the field of model theory in mathematical logic.

> The logical theory which is called the *calculus of (binary) relations*, and which will constitute the subject of this paper has had a strange and rather capricious line of historical development. Although some scattered remarks regarding the concept of relations are to be found already in the writings of medieval logicians, it is only within the last hundred years that this topic has become the subject of systematic investigation. The first beginnings of the contemporary theory of relations are to be found in the writings of A. De Morgan, who carried out extensive investigations in this domain in the fifties of the

---

[4]Brent, J.: *Charles Sanders Peirce: A Life, Revised and Enlarged Edition.* Bloomington, Indiana University Press, 1998

[5]Brent, J.: The Singular Experience of the Peirce Biographer. In www.iupui.edu/ arisbe/menu/library/aboutcsp/brent/singular.htm

[6]Anellis, I.: How Peircean was the "Fregean Revolution" in Logic? In iphras.ru/uplfile/logic/log18/LI-18_Anellis.pdf

[7]Hintikka, J.: The Place of C.S. Peirce in the History of Logical Theory. In *The Rule of Reason: The Philosophy of Charles Sanders Peirce.* Ed. Brunning, J. and Forster, P. Bloomington, Indiana University Press, 1997

[8]Tarski, A.: On the Calculus of Relations. *Journal of Symbolic Logic,* 6: 73–74, 1941.

Nineteenth Century. De Morgan clearly realized the inadequacy of traditional logic for the expression and justification, not merely of the more intricate arguments of mathematics and the sciences, but even of simple arguments occurring in every-day life; witness his famous aphorism, that all the logic of Aristotle does not permit us, from the fact that a horse is an animal, to conclude that the head of a horse is the head of an animal.[9] In his effort to break the bonds of traditional logic and to expand the limits of logical inquiry, he directed his attention to the general concept of relations and fully recognized its significance. Nevertheless, De Morgan cannot be regarded as the creator of the modern theory of relations, since he did not possess an adequate apparatus for treating the subject in which he was interested, and was apparently unable to create such an apparatus. His investigations on relations show a lack of clarity and rigor which perhaps accounts for the neglect into which they fell in the following years.

The title of creator of the theory of relations was reserved for C. S. Peirce. In several papers published between 1870 and 1882, he introduced and made precise all the fundamental concepts of the theory of relations and formulated and established its fundamental laws. Thus Peirce laid the foundation for the theory of relations as a deductive discipline; moreover he initiated the discussion of more profound problems in this domain. In particular, his investigations made it clear that a large part of the theory of relations can be presented as a calculus which is formally much like the calculus of classes developed by G. Boole and W. S. Jevons, but which greatly exceeds it in richness of expression and is therefore incomparably more interesting from the deductive point of view.

Peirce's work was continued and extended in a very thorough and systematic way by E. Schröder...

It is therefore rather amazing that Peirce and Schröder did not have many followers.

As Isaac Newton stood on the shoulders of Galileo and Kepler, Charles Peirce stood on the shoulders of Boole and De Morgan.

Alan James Iliff
North Park University
Chicago, April 2018

---

[9] Peirce's solution to De Morgan's aphorism is in Section 2.2 of this book.

# Original Preface (1992)

The purpose of this book is to discuss the work of Charles Sanders Peirce, specifically his contributions to mathematical logic and especially the work that has had a lasting influence. The historical method used distinguishes two stages, as explained in Chapter 1, Section 1.5: Stage One reconstructs Peirce's work in mathematical logic using present-day notions; Stage Two inquires into the nature and causes of the differences between Peirce's ideas or techniques and contemporary ones. This leads us to the question of why Peirce has not been better recognized for the originality and importance of his work or for the extent of its influence on Leopold Löwenheim and Thoralf Skolem. The major, technical part of this book—Chapters 2, 3, and 6—is concerned with these matters.

The methods used in treating some of Peirce's contributions to mathematics education and philosophy will be less exact, as he himself might have expected, since for him mathematics was the most precise and exact of all the sciences. In these two areas Peirce had original and important ideas, regardless of how much or how little influence his ideas have had.

A discussion of Peirce is essential in order to establish that the model-theoretic or semantic side of logic had its own development which goes back to Augustus De Morgan's work and which is separate from the development of the deduction-theoretic or syntactic side of logic. By 1885 Peirce had laid down the main elements of the framework of quantificational logic within which the work of Löwenheim (1915) and Skolem (1920; 1923) took place. There needs to be a revision of the history of logic, one that will put into balance the account of the model-theoretic line of development with that of the deduction-theoretic line. The significance of this book therefore depends in part on the extent to which it shall contribute to such a revision of the history of logic.

There are certain standard forms of reference used in Peirce scholarship. The editors of the eight volumes of *Collected Papers of Charles Sanders Peirce* numbered the paragraphs in each volume. Peirce scholars make reference to these volumes by means of the form, "CP [volume].[paragraph(s)]"; for example, "CP 3.328–358" denotes Peirce's 1883 paper on the logic of relatives. In this book all references are given by page number; thus,

(Peirce, 1933a, pp. 195–209) would refer to the same 1883 paper in volume 3 of the *Collected Papers.*

Each of Peirce's publications is given an authoritative identifying number in *A Comprehensive Bibliography of the Published Works of Charles Sanders Peirce* (Ketner, 1986). For example, "P 268d" denotes his 1883 paper. These identifying numbers are used in *Writings of Charles S. Peirce*, as discussed in the following paragraph and in Appendix A.

Each of Peirce's *unpublished* manuscripts is being given a chronological number by the Peirce Edition Project at Indianapolis. These enumerations are given in *Writings of Charles S. Peirce: A Chronological Edition*, of which volumes one through six and volume eight are in print. An earlier catalog of Peirce's unpublished manuscripts was given in (Robin, 1967; 1971), but the state of Peirce scholarship at that time precluded the establishment of an exact chronology. In this book, all references to manuscripts are made to published sources, and the nature of the manuscript is indicated in the text. Peirce scholars make reference to these volumes by means of the form, "W [volume]:[page(s)]"; for example, "W 4:453–466" refers to Peirce's 1883 paper. In this book such a reference to volume 4 of *Writings* would be made as (Peirce, 1986b, pp. 453–466).

For the convenience of the reader, whenever it is possible and useful, references will be given both to *Collected Papers* and to *Writings*. Thus (Peirce, 1933a, pp. 195–209; 1986b, pp. 453–466) denotes his 1883 paper, first in the *Collected Papers* and second in the *Writings*. Two appendices are provided for the reader's convenience: Appendix A is an annotated bibliography of Peirce's logic and mathematics papers which are most important to this book, and Appendix B is a chronology of Peirce and his concept of pragmatism.

Charles S. Peirce is generally regarded today as one of the most outstanding philosophers in American history, and especially as the inventor of pragmatism. Nevertheless, he also discovered several of the most important concepts of twentieth century mathematical logic, including the quantifiers, the interpretation of first-order logic by means of relations, and the concept of logical consequence. There is very little general knowledge of Peirce's influence on the development of mathematical logic and almost total ignorance of the details of that influence. The main technical results of this book establish that Peirce laid down the main elements of a framework for the model-theoretic line of development in mathematical logic.

Chapter 1 serves as an introduction in which are given: an explanation of the subject and spirit of this book, a brief life of Peirce with some reference to his works, a preview of Peirce's views on the relation between mathematics and logic, and a description of the historical method used herein.

Chapters 2, 3, and 6 belong strictly to the history of mathematical logic.

Chapters 2 and 3 contain the main technical content—an analysis of material from Peirce's principal technical papers in mathematical logic. Using the work of Boole and De Morgan, Peirce developed an algebra of binary relations and classes in his paper of 1870; this system is the subject of Chapter 2. In Chapter 3, an account is given of how Peirce discovered the quantifiers $\Pi$ and $\Sigma$ (known today as $\forall$ and $\exists$) and how he wrote of this discovery in his papers of 1883 and 1885. By this we really mean that he discovered first-order logic when he realized that there had to be individual variables, which he called indices, in order to express quantification.

In Chapter 4, Peirce's view of the relation between mathematics and logic is discussed and contrasted with the logicist program of Gottlob Frege.

In Chapter 5, Peirce's concept of pragmatism is explained and related to the ancient and present-day usages of words derived from the Greek word, *pragma.* A history of his development of the concept is given: his discussions with a group of associates in Cambridge, his announcement of the pragmatic maxim in 1872, his further refinement of the meaning of the maxim, and his attempts to distinguish his own pragmatism as a response to the appropriation of the word by others.

In Chapter 6, Peirce's place in the history of logic and why his work is not better known are documented and explained. We trace the continuing chain of influences that Peirce's ideas had on certain mathematical logicians, especially Leopold Löwenheim and Thoralf Skolem, and we examine the reasons why Peirce has not been better recognized for his contributions as established in Chapters 2 and 3. This leads to the larger question of why the deduction-theoretic line of development is well-known and the model-theoretic line of development—at least that part of it prior to 1936—is virtually ignored in the accepted history of logic. Peirce's papers were mathematically very difficult for his potential audience because they involved model-theoretic ideas but did not use a clear set-theoretic semantics; moreover, his style of exposition is difficult even for mathematicians to follow.

In the concluding Chapter 7, we sketch some developments of Peirce's ideas in computer science and education. Peirce's work influenced computer science in at least two major ways: Edgar Codd's invention of relational data bases and the development of Prolog through the work of logicians and computer scientists. We then relate Peirce's life and thought to some aspects of mathematics education.

# Chapter 1

# Introduction

Peirce stated in the initial lecture of his series on British logicians, which he delivered at Harvard in 1869:

> The history of logic is not altogether without an interest as a branch of history. For so far as the logic of an age adequately represents the methods of thought of that age, its history is a history of the human mind in its most essential relation,—that is to say with reference to its power of investigating the truth. But the chief value of the study of historical philosophy is that it disciplines the mind to regard philosophy in a cold and scientific eye and not with passion as though philosophers were contestants. (Peirce, 1984, p. 310)

And Leibniz said of history,

> Its use is not just that History may give everyone his due and that others may look forward to similar praise, but also that the art of discovery be promoted and its method known through illustrious examples. (Weil, 1978, p. 227)

## 1.1 The inquiry of this book

The inquiry that resulted in this book began in doubts regarding the existing accounts of Charles S. Peirce's contributions to mathematical logic. The inquiry was conducted in the spirit of B. L. van der Waerden, as stated at the beginning of his work, *Science Awakening*:

> It is the intention to make this book scientific... in the sense that it is to be based on a study of the sources and that its conclusions are to be supported by arguments, so as to enable the reader to judge the conclusions for himself.

> The naive reader may take the use of such a method for granted.
> But—how often has it been sinned against! How frequently it
> happens that books on the history of mathematics copy their
> assertions uncritically from other books, without consulting the
> sources! How many fairy tales circulate as "universally known
> truths"! (van der Waerden, 1961, p. 6)

Naturally, then, this book is based on the publications of Charles S. Peirce;
the manuscripts which Peirce himself did not publish will occasionally be
used to aid in understanding his published work. Peirce expressed a view
similar to van der Waerden's in the manuscript for another lecture in his
1869 series on British logicians:

> A theory of science which is thus founded on the history of
> science in a truly scientific spirit and by a genuine inductive
> method and which does not merely make use of facts of scientific
> history to support a theory which has really been derived not
> from these but from a general philosophical doctrine of a meta-
> physical origin, must be true to the grand features of scientific
> progress, to all those characters of scientific investigation which
> leave their mark upon its history... But scientific progress is to
> a large extent public and belongs to the community of scientific
> men of the same department, its conclusions are unanimous,
> its interpretations of nature are no private interpretations, and
> so much must always be published to the world as will suf-
> fice to enable the world to adopt the individual investigator's
> conclusions. (Peirce, 1984, p. 339)

In examining Peirce's contributions to mathematical logic, we shall see
both that the accepted history of logic contains its share of "fairy tales"
circulating as "universally known truths", and that it is based on a theory
derived "from a general philosophical doctrine of a metaphysical origin"
rather than from the facts of written documents and the verifiable influence
of ideas. One example is a strange 'history' in which today's mathematical
logic is the result of a line of development initiated by Gottlob Frege.

## 1.2   A brief life of Peirce

Charles Sanders Peirce was born on September 10, 1839, the son of Benjamin Peirce, who was Professor of Mathematics and Astronomy at Harvard,
and the most eminent mathematician in America. From his earliest childhood Peirce was exposed to great scientists and their ideas as they visited
his parents' home. As a child, he worked on what was at that time graduate level chemistry and mathematics. He entered Harvard in 1855 and
graduated in 1859. In the spring of 1861 he entered the Lawrence Scientific

School where he performed brilliantly and took an Sc.B. degree *summa cum laude* in 1863. (Fisch, 1982, p. xx)

Peirce was, first of all, a research scientist with the U.S. Coast Survey, which later became the U.S. Coast and Geodetic Survey, and now comprises both the National Oceanic and Atmospheric Administration (NOAA) and the National Aeronautics and Space Administration (NASA). Second, he was an academic of sorts, delivering lectures at Harvard and elsewhere, publishing technical papers in mathematics and logic, and for a few years teaching logic and directing graduate students at Johns Hopkins University. And third, he was a 'philosopher' in the ancient sense, carrying out his inquiries through some published papers and an immense body of unpublished manuscripts.

Peirce's career at the Survey lasted from 1861 to 1891. His accomplishments in this capacity were enormous. He planned and participated in expeditions to survey the United States, and at Harvard Observatory he did outstanding work in astronomy. He helped to plan and participated in international cooperative scientific expeditions to observe solar eclipses. Peirce's later work concerned variations in the strength of gravity over the earth's surface, or geodesy, and he carried out experiments and improved the apparatus for measuring the force of gravity, one of the most important physical constants. He invented the quincuncial projection map, which is still in use today because the great circle routes of global air and sea travel are plotted nearly as straight lines on such a conformal map. He proposed that a standard of length be defined by means of a particular wavelength of light, which has become the standard definition. In his only published book, *Photometric Researches*, he made the first determination of the shape of the Milky Way galaxy (1986a, pp. 382–493). In some of these activities Peirce cooperated in the first truly international scientific projects. Political harassment of the Survey and personal rivalries forced Peirce to resign in 1891. By that time, he had earned a reputation as one of America's leading men of science, and it was through this scientific work that he was best known to his contemporaries.

Less well-known to his contemporaries was Peirce's career as a research mathematician and logician. He gave many lectures on mathematics and logic. He published a number of important papers that laid the groundwork for the model-theoretic line of development in mathematical logic. In 1876 Johns Hopkins University was founded, and from 1879 to 1884 Peirce was a part-time lecturer in logic there while he continued to work for the Survey. This brought him into contact with a remarkable faculty and brilliant students. During this period he wrote most of his important technical papers in mathematics and logic. In 1883 he edited *Studies in Logic*, a book of his own and his students' papers; this book demonstrated the quality of his teaching through the accomplishments of his students. The treachery of a rival at the Coast Survey brought about his dismissal from

Johns Hopkins, although the trustees conspired to make the firing appear to be a reorganization of faculty duties. (Houser, 1986, pp. lxii–lxv)

But hardly known at all to his contemporaries was Peirce's career as a philosopher—not an academic philosophy professor, but "a wise man, a student of all things in the sky and below the earth" (Plato, 1975, p. 22). During the 1860s Peirce gave several courses of lectures at Harvard on the history of logic and philosophy. In the early 1870s Peirce and others, including William James, formed a philosophical discussion group in Cambridge, which they called the "Metaphysical Club". As a result of these discussions and of his own philosophical work, Peirce published his pragmatic maxim in the second of an unfinished series of articles about scientific method that appeared in the *Popular Science Monthly* for 1877–78—a series of articles written not for academics but for the educated public. When Peirce resigned from the Survey in 1891, he moved to Milford, Pennsylvania, where he had used an inheritance in 1887 to buy a house which he called "Arisbe"[1]. He lived his last twenty-three years in Milford, supporting himself on his writings, his dwindling inheritance, occasional lectures, a bit of engineering consulting, and the charity of friends. Although he wrote tens of thousands of manuscript pages on philosophy, little of his later work was published during his lifetime; most of it remains only in the form of fragmentary manuscripts. Pragmatism as a philosophy became well-known in his last years, but the word became popular without carrying much of Peirce's thought or writing along with it. He died destitute on April 19, 1914.

The most important point to be made here is that this unusual combination of careers was to some extent deliberate: Peirce used the practice of science to cultivate his own pragmatic habit of mind and to protect himself from the impractical, ivory tower introspection of purely academic philosophy. He desired contact with any and all vital inquiries that could be treated using a generalization of the scientific method—whether about accepted sciences, about law, or about any other such matters—whether in an academic setting or elsewhere. Peirce certainly suffered himself greatly because he did not suffer fools gladly. During his later years he attempted a number of publishing projects, including treatises on logic and a series of mathematics textbooks for grade schools, but unfortunately none of these were completed in his lifetime.

According to Charles Hartshorne,[2] who knew Peirce's widow, she got Harvard University to buy Peirce's manuscripts from her shortly after he died. Later, Hartshorne was drafted, with Paul Weiss volunteering to help, into the project of editing Peirce's papers. But there was an overwhelming amount of material, and the editors were limited by their training to

---

[1] One of the cities mentioned by Homer in the *Iliad* as an ally of Troy. The significance of the name will be discussed in Section 6.2 below.

[2] Material in this paragraph is from a September 1989 conversation between Hartshorne and the author.

academic philosophy, which they chose to emphasize in the six volumes (out of a projected twelve) of the *Collected Papers of Charles Sanders Peirce* that they brought out between 1931 and 1935. Hartshorne says that they expected others to edit the scientific and mathematical work later. Nevertheless, they did assemble Peirce's most important mathematical and logical publications in volume 3 (Peirce, 1933a), and they made a selection from his unpublished writings on mathematics and logic for volume 4 (1933b). (Hartshorne, 1989, personal communication)

In 1958, Arthur Burks finished editing the *Collected Papers* by bringing out volumes 7 and 8 (1958a; 1958b). In 1976 Carolyn Eisele edited the manuscripts of the mathematics textbooks written by Peirce, together with other papers and letters on mathematics and mathematical philosophy; these were published as *The New Elements of Mathematics*, a title Peirce had used for the textbooks (1976a; 1976b; 1976c; 1976d). In 1982, the Peirce Project, under the direction of Max Fisch and other Peirce scholars at Indiana and Purdue Universities in Indianapolis, brought out the first volume of a projected thirty volume edition of Peirce's writings, to be based on chronological principles. In 1985 Carolyn Eisele edited some of Peirce's manuscripts on the history of science as, *Historical Perspectives on Peirce's Logic of Science: A History of Science* (1985).

Ironically, although Peirce was best known to his contemporaries as a scientist and least as a philosopher, he is generally regarded today as one of the leading philosophers in American history. In the recovery of Peirce's reputation, the fame of his philosophy of pragmatism has preceded a proper assessment of his place in the history of science, mathematics, and logic. Peirce discovered several of the most important concepts of mathematical logic, including the quantifiers, the interpretation of a first-order language by means of relations, and the concept of logical consequence. The purpose of the technical part of this book is to examine some of the steps whereby Peirce laid down the main elements of quantificational logic. Historically, this work is of especial importance because it provided the main elements of the framework of quantificational logic within which the 1915 paper of Leopold Löwenheim and the 1920 and 1923 papers of Thoralf Skolem were written.

Authoritative material on Peirce's life and intellectual development is given in the introductory material to the volumes of *Writings of Charles S. Peirce*. Much of this work depended on research by Max Fisch, and the volume of his essays is very useful (1986b). Ewald's review of the first four volumes of *Writings* contains a good brief account of Peirce and his work (1990), and so does Ketner's introduction to some selections by Peirce (1987). Information on Peirce's development of pragmatism may be found in (Fisch, 1986a; 1986b), (Thayer, 1981a), and (P. Wiener, 1949; 1973). No clear and comprehensive account of Peirce's contributions to mathematical logic is available, but certain facts and details can be found in (Beth, 1968), (Bochenski, 1970), (Brady, 1990), (Church, 1956), (Curry, 1963), (Goldfarb, 1971; 1979), (Kneale and Kneale, 1962), (Merrill, 1984), (Moore, 1977; 1988), (Putnam, 1982), and (van Heijenoort, 1977b). Putnam, van Heijenoort, and some others have suggested that there is a line of development including Boole, De Morgan, Peirce, Schröder, Löwenheim, and Skolem, but they have not maintained and developed the leading idea that Peirce laid down the main elements of a framework for first- and second-order logic as is done in (Brady, 1990) and in this book.

## 1.3   The Peircean Knot

Many attempts have been made to find a unity underlying Peirce's work. Peirce himself gave a number of tantalizing hints that he had developed one great system, but the extensive written works that would have been necessary to expound such a system did not appear. Peirce's publications were on so many subjects and in so many fields that it would take an unusual individual to master all of them, and then only after many years of study. His unpublished papers confront the scholar with multiple, alternate drafts of some of the publications, of lectures delivered and undelivered, of chapters for large works on philosophy and logic that were never published, and so on. The enormous range of Peirce's thought invites comparison with only one other person—Leibniz.

In a review of the first four volumes of *Writings of Charles S. Peirce*, W. B. Ewald states:

> Near the end of his strange and beleaguered life, shortly after
> the death of his friend William James, C. S. Peirce wrote
> despairingly: "Who could be of a nature so different from his
> as I? He so concrete, so living; I a mere table of contents, so
> abstract, a very snarl of twine."[3] Anybody who has struggled
> to understand Peirce's thought—or who has turned to Peirce
> from tidier thinkers like James or Frege—will feel at once
> the justice of his self-appraisal. Always trying to squeeze
> his thoughts into a system, he never managed to make them
> fit. They keep brimming over, refusing to stay within their
> assigned boundaries. The insights and suggestions and projects
> come tumbling out of him in an exuberant cascade: logic
> machines, an indeterminate physics, a differential calculus based
> on infinitesimals, pungent observations on scientific method,
> and a steady stream of articles on logic and the foundations of
> mathematics. (Ewald, 1990, p. 599)

Ewald goes on to modify Peirce's own metaphor of a "snarl of twine" into an actual "knot":

---

[3]Ewald's quotation is from (Peirce, 1935, p. 131).

Peirce's logical and scientific research reacted on his philosophy—on his conception of scientific method, language, truth, reality. And his work in philosophy and the history of science in turn suggested new lines of research in logic, just as did his work in the natural sciences. As these papers show, no one of these activities had priority in Peirce's thinking, and indeed his conception of philosophy as based on a public, non-psychologistic study of scientific reasoning was bound to tie logic, natural science and philosophy into a tight knot. (Ewald, 1990, p. 600)

One who desires to comprehend this enormous range of thought might well consider the story of Alexander and the Gordian Knot. Plutarch actually recounted two versions:

...after he had taken the city of Gordium, reputed to have been the home of the ancient Midas, he saw the much-talked-of waggon [*sic*] bound fast to its yoke with bark of the cornel-tree, and heard a story confidently told about it by the Barbarians, to the effect that whosoever loosed the fastening was destined to become king of the whole world. Well, then, most writers say that since the fastenings had their ends concealed, and were intertwined many times in crooked coils, Alexander was at a loss how to proceed, and finally loosened the knot by cutting it through with his sword, and that when it was thus smitten many ends were to be seen. But Aristobulus says that he undid it very easily, by simply taking out the so-called "hestor," or *pin*, of the waggon-pole, by which the yoke-fastening was held together, and then drawing away the yoke. (Plutarch, 1919, p. 273)

Some investigators have tried to cut the Peircean Knot—separating his logic from his metaphysics, his mathematics from his logic, his semeiotics from his mathematics, his scientific achievements from his pragmatism, and so on. Some have juxtaposed short passages, taken out of context from Peirce's vast writings, in order to support their own viewpoint—usually one of a metaphysical origin rather than one based on his mathematical and logical papers—as if they might reassemble selected pieces of the knot in a manner to their own liking.

It is a major but implicit theme in this book that rather than cutting this Peircean Knot, one can leave it intact. Nevertheless, this can only be done by comprehending and reconstructing his mathematical ideas and his mathematical logic—taking out the pin, as it were. For Peirce, mathematics was the most abstract of all the sciences, and he maintained that the logic needed in science and philosophy must be based on the reasoning used by mathematicians.

# 1.4 Peirce, mathematics, and logic

In Chapter 4 we shall consider how Peirce viewed the relation between mathematics and logic and how different is his view from Frege's. Since we are about to analyze in detail a number of passages from Peirce's writings on logic, it is appropriate to provide a preview: Peirce maintained that mathematics was the science with the most effective reasoning methods and that a logic useful to philosophy and science must therefore be based on the reasoning used by mathematicians. In other words, in a certain sense, for Peirce logic depends on mathematics—a claim which is in vivid contrast to the logicist claim that mathematics is a part of, and must be founded on, logic.

At the beginning of Peirce's 1896 review of Ernst Schröder's *Exact Logic*, he gave his most thorough published statement about the relation between mathematics and logic. He began by observing that mathematics is, "the only science in which there has never been a prolonged dispute concerning the proper objects of that science." (1933a, p. 268) From this empirical observation, Peirce inferred that mathematicians have developed the most advanced reasoning method and that this method must be taken as the ideal to which all scientific method should be compared. The following statements by Peirce will be discussed more fully in Section 4.2 below:

> ...the reason for this immunity of mathematics ...arises from the fact that the objects which the mathematician observes and to which his conclusions relate are objects of his mind's own creation. ..., it is so easy to repeat the inductions upon new instances, ...that when attention has once been directed to a process of reasoning suspected of being faulty, it is soon put beyond all dispute either as correct or as incorrect.

> ...considering the immense amount of disputation there has always been concerning the doctrines of logic, and especially concerning those which would otherwise be applicable to settle disputes concerning the accuracy of reasonings in metaphysics, the safest way is to appeal for our logical principles to the science of mathematics, where error can only long go unexploded on condition of its not being suspected.

> This double assertion, first, that logic ought to draw upon mathematics for control of disputed principles, and second that ontological philosophy ought in like manner to draw upon logic, is a case under a general assertion ...that the sciences may be arranged in a series with reference to the abstractness of their objects; and that each science draws regulating principles from those superior to it in abstractness, while drawing data for its inductions from the sciences inferior to it in abstractness...

MATHEMATICS

Philosophy $\left\{ \begin{array}{l} \text{Logic} \\ \text{Metaphysics} \end{array} \right.$

| Science of Time | Geometry | |
|---|---|---|
| Nomological Psychics | Nomological Physics | $\left\{ \begin{array}{l} \text{Molar} \\ \text{Molecular} \\ \text{Ethereal} \end{array} \right.$ |
| Classificatory Psychics | Classificatory Physics | $\left\{ \begin{array}{l} \text{Chemistry} \\ \text{Biology, or the} \\ \text{chemistry of} \\ \text{protoplasms} \end{array} \right.$ |
| Descriptive Psychics | Descriptive Physics | |

PRACTICAL SCIENCE

Table 1.1: Mathematics, philosophy, and the sciences
(Peirce, 1933a, p. 270)

Mathematics is the most abstract of all the sciences. For it makes no external observations, nor asserts anything as a real fact. When the mathematician deals with facts, they become for him mere "hypotheses"; for with their truth he refuses to concern himself. The whole science of mathematics is a science of hypotheses; so that nothing could be more completely abstracted from concrete reality. Philosophy is not quite so abstract. ...philosophy, in the strictest sense, confines itself to such observations as *must* be open to every intelligence which can learn from experience. ...logic is much more abstract even than metaphysics. For it does not concern itself with any facts not implied in the supposition of an unlimited applicability of language. (Peirce, 1933a, pp. 269–271)

Thus, for Peirce mathematics is the most abstract science because it deals exclusively in hypotheses; nevertheless, it is still an experimental science because the mathematician observes examples in thought and makes inductions from them. Even though the mathematician's thought is fallible, it is so easy to repeat these thought experiments and to make new ones, that mathematics has provided the opportunity to develop the most perfect paradigm for the scientific method. And thus, logic should be based on mathematics, or at least, on mathematical thought and mathematical reasoning.

19

In these statements, Peirce has told us in effect how the Peircean Knot was tied. That is to say, he implied that his schema—in which all of the sciences depend on logic and logic depends on mathematics—indicated the very organization which he strove to give to his own intellect and to his work.

## 1.5  Peirce's technical work and the method of this book

Peirce's first important paper in mathematical logic was published in 1870; it contained his two-sorted algebra of classes and binary relations. His other four principal papers in mathematical logic were published in 1880, 1882, 1883, and 1885. (See Appendix A for an annotated bibliography of his logic and mathematics papers that are most important to this book.)

Two different systems were developed by Peirce in these papers: the first was his 1870 algebra of binary relations and classes; the second was his first- and second-order logic with quantifiers, logical connectives, and constants and variables for individuals and n-ary relations, which was developed in the papers of 1883 and 1885. In the next two chapters we make two contributions toward the clarification of Peirce's work on these two systems: In Chapter 2 we provide an introduction to his 1870 paper and give an analysis of two proofs in that paper; in Chapter 3 we give an account of Peirce's ideas, especially abstract algebraic ones, which led him to the discovery of the quantifiers and his subsequent development of first- and second-order logic.

The first system is important both because it gave rise to the second and because it had a major influence on Schröder. The second system, a general framework of first- and second-order logic, was further developed by Schröder and passed on through his work to Löwenheim and Skolem. In Chapter 6 we discuss how first-order logic, which Peirce called "first-intentional logic", retained its integrity in Schröder's work and provided the framework for Löwenheim's Theorem (1915) and the Löwenheim-Skolem Theorem (Skolem, 1920; 1923).

A number of Peirce's results—his existential graphs, his definition of a finite set, his development of lattices in the 1880 paper, Peirce's Law, etc.—are not discussed in this book because they did not play much of a role in the Peirce–Schröder–Löwenheim–Skolem line of development. There are other possible avenues of Peirce's influence on logic which need to be investigated, such as whether and to what extent his work influenced Bertrand Russell or Alfred Tarski, but the story in Chapters 2, 3, and 6 does not depend on such investigations. There is also a question, discussed in Section 3.4.2, whether Peirce knew of Frege's *Begriffsschrift* (1879); but even if he did, his own work was developed by means of a system of ideas quite different from Frege's.

There are two separate lines of development in mathematical logic: the two lines, which we shall call "deduction-theoretic" and "model-theoretic", can only be understood in retrospect, after the development of rigorous concepts of syntax and semantics. On the one hand, the tradition of Frege, Peano, Russell, and Whitehead is essentially deduction-theoretic; on the other hand, the tradition of Boole, De Morgan, Peirce, Schröder, Löwenheim, and Skolem is essentially model-theoretic. In his paper of 1885, Peirce laid down the main elements of the framework of quantificational logic within which Löwenheim and Skolem worked. To conflate the two lines of development, referring to the tradition of Peirce and Schröder as the "algebra of logic", is to ignore the culmination of this early model-theoretic line in the theorems of Löwenheim and Skolem, which were a natural outgrowth of Peirce's work.

So why is Peirce not better known for his development of first-order logic? That will be the subject for Chapter 6. Suffice it for now to say that his papers were mathematically too difficult for most of his potential audience and that his style of exposition was difficult even for mathematicians to follow. This difficult style, together with Peirce's lack of the set-theoretic concepts familiar to present-day mathematicians, makes an account of these papers something more than mere exposition. Indeed, it requires that various passages be deciphered and that the contents be reconstructed using set-theoretic concepts. Moreover, this decipherment and reconstruction must be carried out with a continual awareness of the dangers of anachronism. As André Weil said in his talk, 'History of Mathematics: Why and How':

> How much mathematical knowledge should one possess in order to deal with mathematical history? According to some, little more is required than what was known to the authors one plans to write about; some go so far as to say that the less one knows, the better one is prepared to read those authors with an open mind and avoid anachronisms. Actually the opposite is true. An understanding in depth of the mathematics of any given period is hardly ever to be achieved without knowledge extending far beyond its ostensible subject-matter. More often than not, what makes it interesting is precisely the early occurrence of concepts and methods destined to emerge only later into the conscious mind of mathematicians; the historian's task is to disengage them and trace their influence or lack of influence on subsequent developments. Anachronism consists in attributing to an author such conscious knowledge as he never possessed... (Weil, 1978, pp. 231–232)

Nevertheless, if we already know what we should be looking for, we will be able to recognize it when and where we find it.

It has been all too easy for certain shortsighted investigators to claim that Peirce was confused or mistaken where he was actually lacking— or even intentionally avoiding—some concept taken for granted by such investigators. And conversely, other investigators have attributed to Peirce "such conscious knowledge as he never possessed".

We need to distinguish two stages of analysis in reading Peirce's papers: Stage One consists of the determination of objective mathematical content in terms of present-day concepts, especially those of set theory; Stage Two consists of an attempt to reconstruct the concepts Peirce used in his arguments and definitions, and especially to avoid the anachronistic fallacies mentioned in the previous quotation. Stage One provides an answer in today's terms to the question of what Peirce discovered or invented; Stage Two proposes answers to the questions of how he did it, why he did it the way he did, and why he did not do more or otherwise than he did. A correct Stage One analysis still leaves gaps in our understanding of Peirce's arguments, because wherever his reasoning is sufficiently different from modern reasoning, only a Stage Two analysis can explain these gaps. Some of his ideas about metaphysics steered him away from the set-theoretic viewpoint that we take for granted today, especially from the viewpoint that the relation 'element of' needs to be taken as fundamental and from the viewpoint that the empty set is a legitimate set.

# Chapter 2

# Binary Relations and Classes

At the beginning of his 1870 paper on the logic of relatives, Peirce stated:

> Relative terms usually receive some slight treatment in works upon logic, but the only considerable investigation into the formal laws which govern them is contained in a valuable paper by Mr. De Morgan in the tenth volume of the *Cambridge Philosophical Transactions*. He there uses a convenient algebraic notation... This system still leaves something to be desired. Moreover, Boole's logical algebra has such singular beauty, so far as it goes, that it is interesting to inquire whether it cannot be extended over the whole realm of formal logic, instead of being restricted to that simplest and least useful part of the subject, the logic of absolute terms, which, when he wrote, was the only formal logic known. The object of this paper is to show that an affirmative answer can be given to this question. I think there can be no doubt that a calculus, or art of drawing inferences, based upon the notation I am to describe, would be perfectly possible and even practically useful in some difficult cases, and particularly in the investigation of logic. I regret that I am not in a situation to be able to perform this labor, but the account here given of the notation itself will afford the ground of a judgment concerning its probable utility. (Peirce, 1933a, p. 27–28; 1984, pp. 359–360)

As background for an understanding of this paper by Peirce—and indeed the whole body of his work—it is necessary to describe certain work of Boole and De Morgan. Next we discuss some of Peirce's work which led to the 1870 paper but which he did not publish. After this preparation, we consider a part of the 1870 paper itself. At the end of the previous chapter it was explained that our historical analysis proceeds in two stages. As a remark on terminology, we note that for a Stage One account of the content of this 1870 paper, we should interpret Peirce's use of "relative" to mean "relation" (usually, binary relation) in the present-day sense. Actually, it was not until Peirce developed the matrix representation of a relation in 1883 that he gained the advantages of today's conception of a relation as a set of ordered pairs.

## 2.1 The work of Boole and De Morgan

As Peirce said at the beginning of the 1870 paper, quoted above, he had combined certain work of Boole and De Morgan. In brief, he was referring to the Boolean algebra of classes and the relational operations of De Morgan. Boole introduced his algebra in his 1847 monograph, *A Mathematical Analysis of Logic: Being an Essay Towards a Calculus of Deductive Reasoning*. It is important to note what Boole meant by the title:

> We might justly assign it as the definitive character of a true Calculus, that it is a method resting upon the employment of Symbols, whose laws of combination are known and general, and whose results admit of a consistent interpretation. (Boole, 1847, p. 4)
>
> A logical proposition is, according to the method of this Essay, expressible by an equation the form of which determines the rules of conversion and of transformation, to which the given proposition is subject...
>
> The premises of a syllogism being expressed by equations, the elimination of a common symbol between them leads to a third equation which expresses the conclusion, this conclusion being always the most general possible... (p. 8)

In order to understand how Peirce modified Boole's program, it is necessary to understand that Boole intended his program to use an equational calculus of classes for the expression of statements and valid inferences in syllogistic logic. Boole also reinterpreted his class algebra as an algebra of propositions (1847, pp. 48–49). In this sense of 'mathematical logic', the symbolism and techniques of mathematics were to be applied to the traditional logic of syllogism. We shall discuss in Chapter 4 how Peirce was also concerned with the other sense of 'mathematical logic'—the analysis of that kind of logic especially used by mathematicians in their reasoning.

Boole succeeded in expressing some syllogistic arguments through substitutions of equations, as in this example of the universal syllogism, Barbara (p. 34):

$$\text{All } Y\text{s are } X\text{s} \quad y(1-x) = 0 \quad \text{or} \quad (1-x)y = 0$$
$$\text{All } Z\text{s are } Y\text{s} \quad z(1-y) = 0 \quad \text{or} \quad zy - z = 0.$$

Applying the substitution form (p. 32),

$$ay + b = 0 \text{ and } a'y + b' = 0 \text{ yields } ab' - a'b = 0,$$

and using

$$a = (1-x), b = 0, a' = z \text{ and } b' = -z,$$

he got

$$(1-x)(-z) - z(0) = -z + xz$$
$$= zx - z$$
$$= 0$$
$$= z - zx$$
$$= z(1-x)$$

$$\therefore \text{ All } Z\text{s are } X\text{s. (p. 34)}$$

But Boole's equational calculus could not express the non-emptiness of a class; that is, he could not find an equation that expressed $x \neq 0$. Hence, he was not able to express the particular proposition adequately, although he tried to use an indefinite but non-empty class symbol, $v$:

To express the Proposition, Some $X$s are $Y$s.

$$v = xy \text{ (p. 21)}$$

In 1854 Boole published *An Investigation of The Laws of Thought on Which Are Founded the Mathematical Theories of Logic and Probabilities*, which was a greatly expanded version of his logic, with many examples of syllogisms and probabilities. However, from the standpoint of the progress of mathematical logic—in the sense of the logic used in mathematical reasoning and proof—the original contribution was obscured or diluted by this concentration on traditional logic, and the 1847 monograph provides a better account of his work.

In the model-theoretic development of logic, the notion of an interpretation is based on the idea of having a domain of individuals. This contrasts with the approach of Frege, whose universe consisted of all objects and all "functions". The beginning of the model-theoretic line of development in mathematical logic can be seen in De Morgan's paper, 'On the Syllogism: I. On the Structure of the Syllogism' (read in 1846 but published in 1849), where he introduced the notion of a domain of individuals:

Writers on logic, it is true, do not find elbow-room enough in anything less than the whole universe of possible conceptions;

but the universe of a particular assertion or argument may be limited in any matter expressed or understood. And this without limitation or alteration of any one rule of logic...

By not dwelling on this power of making what we may properly (inventing a new technical name) call the *universe* of a proposition, or of a name, matter of express definition, all rules remaining the same, writers on logic deprive themselves of much useful illustration. And, more than this, they give an indefinite negative character to the *contrary*, as Aristotle did when he said that not-man was not the name of anything. Let the universe in question be 'man:' then *Briton* and *alien* are simple contraries; alien has no meaning of definition except not-Briton. But we cannot say that either term is positive or negative, except correlatively. (De Morgan, 1966, p. 2)

It is very important to note well De Morgan's implication that the same logical rules must apply to all universes or models. In *A Mathematical Analysis of Logic*, Boole used the term "Universe":

Let us employ the symbol 1, or unity, to represent the Universe, and let us understand it as comprehending every conceivable class of objects whether actually existing or not... (Boole, 1847, p. 15)

But in *The Laws of Thought*, Boole used the term "universe of discourse" more precisely in De Morgan's way:

Now, whatever may be the extent of the field within which all the objects of our discourse are found, that field may properly be termed the universe of discourse. (Boole, 1854, p. 42)

By contrast, Frege did "not find elbow-room enough in anything less than the whole universe of possible conceptions" (De Morgan, 1966, p. 2)—in his case the universe of all objects and all "functions".

De Morgan's second contribution to the beginning of the model-theoretic line was a mathematical conception of binary relations, which appeared in his paper, 'On the Syllogism IV; and on the Logic of Relations' (read in 1860 but published in 1864):

Any two objects of thought brought together by the mind, and thought together in one act of thought, are *in relation*... Two thoughts cannot be brought together in thought except by a thought: which last thought contains their *relation*. (De Morgan, 1966, p. 218)

I do not use the mathematical symbols of functional relation, $\phi$, $\psi$, &c.: there are more reasons than one why mathematical examples are not well suited for illustration. The most apposite instances are taken from the relations between human beings: among which the *relations* which have almost monopolized the name, those of consanguinity and affinity, are conspicuously convenient, as being in daily use. (p. 220)

Note that De Morgan said he would avoid taking examples from mathematics. One of the tasks in performing a Stage One analysis of Peirce's work lies in resolving the inevitable ambiguities of such relation examples as "lover of", "servant of", "benefactor of", etc.; we must replace them with examples from modern mathematics, especially set-theoretic ones, and we begin to do this in the next section. For now, we continue with some quotations from De Morgan's work:

Let $X..LY$ signify that $X$ is some one of the objects of thought which stand to $Y$ in the relation $L$, or is one of the $L$s of $Y$. Let $X.LY$ signify that $X$ is not any one of the $L$s of $Y$. (De Morgan, 1966, p. 220)

When the predicate is itself the subject of a relation, there may be a composition: thus if $X..L(MY)$, if $X$ be one of the $L$s of one of the $M$s of $Y$, we may think of $X$ as an '$L$ of $M$' of $Y$, expressed by $X..(LM)Y$, or simply by $X..LMY$. (p. 221)

Note the associativity of composition of binary relations in the previous passage. To continue,

We have thus three symbols of compound relation: $LM$, an $L$ of an $M$; $LM'$, an $L$ of every $M$; $L, M$, an $L$ of none but $M$s.

...The *converse* relation of $L$, $L^{-1}$, is defined as usual: if $X..LY$, $Y..L^{-1}X$: if $X$ be one of the $L$s of $Y$, $Y$ is one of the $L^{-1}$s of $X$...

...Relations are assumed to exist between any two terms whatsoever. If $X$ be not any $L$ of $Y$, $X$ is to $Y$ in some not-$L$ relation: let this *contrary* relation be signified by $l$; thus $X.LY$ gives and is given by $X..lY$.

...Contraries of converses are converses...

...Converses of contraries are contraries...

(De Morgan, 1966, pp. 222–223)

Note that by "converse", De Morgan meant what is called today the inverse. Thus he gave a notation capable of expressing many facts about binary relations, but he did not have a calculus in Boole's sense.

## 2.2 Peirce combines their work

Peirce first discussed Boole's work in his 1865 Harvard Lectures on the "Logic of Science". In two of these lectures he gave an exposition of Boole's calculus of logic and its application to probabilities (1976c, pp. 298–327; 1982, pp. 189–204, 223–239). Peirce was not satisfied with Boole's attempt to express the particular syllogism in equational form:

> In Quantity, ordinary language expresses the difference of Universal and Particular. Professor Boole gave as the expression for the particular negative Some $X$ is not $Y$
>
> $$vx = v(1 - y)$$
>
> where $v$ denotes the indefinite class *some*. But the absurdity of this is evident from the fact that by transposing we get
>
> $$vy = v(1 - x)$$
>
> or Some $Y$ is not $X$. But it does not follow from Some $X$ is not $Y$ that Some $Y$ is not $X$. This expression is therefore wrong. The cause of the defect of it is evident. He has represented *some* as being merely an indefinite class. (Peirce, 1976c, pp. 319–320; 1982, pp. 230–231)

Peirce then tried but failed to give a correct expression (1976c, p. 320; 1982, p. 231). As we shall see later in the present section and in Section 2.7 below, this problem was very important to Peirce; he continued to work at it until he solved it by extending Boole's system to incorporate De Morgan's relations.

In 1867 Peirce published his paper, 'On an Improvement in Boole's Calculus of Logic' (1933a, pp. 3–15; 1984, pp. 12–23), in which he proposed the present-day notion of set-theoretic union or inclusive disjunction. Boole took the addition sign of his own system to signify a disjoint union or an exclusive disjunction, which required that the sum of two non-disjoint sets be empty. For example, $a + a = 0$. Peirce distinguished the logical operations (or class operations) from the arithmetic ones by adding commas: union was symbolized by a plus sign followed by a comma, intersection by a comma (instead of juxtaposition), and equality by an equal sign with a comma under it. In his 1867 paper, Peirce explained his improvement to Boole's system in this way:

> Let $a \mathbin{+\!\!\!,} b$ denote all the individuals contained under $a$ and $b$ together. The operation here performed will differ from arithmetical addition in two respects: 1st, that it has reference to identity, not to equality; and 2d, that what is common to $a$ and $b$ is not taken into account twice over, as it would be in

arithmetic. The first of these differences, however, amounts to nothing, inasmuch as the sign of identity would indicate the distinction in which it is founded; and therefore we may say that

$$\text{If No } a \text{ is } b \qquad a \mathbin{\substack{+\\,\cdot}} b \mathbin{\substack{=\\,\cdot}} a + b$$

It is plain that

$$a \mathbin{\substack{+\\,\cdot}} a \mathbin{\substack{=\\,\cdot}} a$$

and also, that the process denoted by $\mathbin{\substack{+\\,\cdot}}$, and which I shall call the process of logical addition, is both commutative and associative. (Peirce, 1933a, pp. 3–4; 1984, pp. 12–13)

From the modern standpoint, Boole's addition operation is objectionable because the operation does not apply to all pairs of classes. (Boole wanted the notation to apply to probabilities besides to classes and propositions; the probabilities of two events can be added to get the probability of the combined event only if the two events are mutually exclusive.) Through Peirce's improvement we can obtain the familiar De Morgan dual distributivity laws. The use of a comma to mark the new kinds of addition, subtraction, and equality was unfortunate, as we shall see in discussing the 1870 paper, because it could cause confusion between the ordinary grammatical use of the comma and these new compound symbols. In his 1867 paper, Peirce also returned to the problem of expressing non-emptiness by means of an equation, but again he failed to do it (1933a, p. 13; 1984, p. 21).

Late in life, Peirce recalled his first encounter with De Morgan's 1860 paper on the logic of relations:

> ...I at once fell to upon it; and before many weeks had come to see in it, as De Morgan had already seen, a brilliant and astonishing illumination of every corner and every vista of logic. ...his was the work of an exploring expedition, which every day comes upon new forms for the study of which leisure is, at the moment, lacking, because additional novelties are coming in and requiring note. He stood indeed like Aladdin (or whoever it was) gazing upon the overwhelming riches of Ali Baba's cave, scarce capable of making a rough inventory of them. (Peirce, 1931, p. 301)

De Morgan had written in his paper:

> ...it is *not* the truth that all inference can be obtained by ordinary syllogism, in which the terms of the conclusion must be terms of the premises. If any one will by such syllogism prove that because every man is an animal, therefore every head of a man is a head of an animal, I shall be ready to—set him another question. (De Morgan, 1966, p. 216)

Peirce's first published reference to this paper of De Morgan's was to quote the last sentence of the previous quotation in a footnote to an 1869 philosophical article (1934, p. 193n1; 1984, p. 245n2). It is now accepted that he began to study De Morgan's paper late in 1868 (Fisch, 1984, p. xxxi; Merrill, 1984, pp. xliii–xlv). The early stages of Peirce's discovery of the algebra of relations can be seen in a passage from his manuscript "Logic Notebook" for November 3, 1868, which is cited by Daniel Merrill (1978, pp. 266–268).

Merrill mentions that Peirce expressed the concept of "lover of woman" by the subscript notation $l_x$; but Peirce was still experimenting with whether this should be "lover of some woman" or "lover of every woman", given that the $l$ stands for the relation "lover of" and the subscript x stands for the class of women.[1] In the case of "some", Peirce found that,

$$l_{x+y} = l_x + l_y,$$

but in the case of "every" he found rather that,

$$l_{x+y} = l_x \cdot l_y.$$

This suggested a law of exponents, and by November 10, 1868: Peirce had changed the "related to every element" notation to a superscript,

$$l^{x+y} = l^x \cdot l^y;$$

he had taken the notation for "related to some element" to be multiplication,

$$l(x + y) = lx + ly;$$

and he had derived a number of results analogous with the laws of exponents as used in arithmetic and algebra (Merrill, 1978, pp. 266–267). For example,

$$l^{x+y} = l^x \cdot l^y$$

can be interpreted as stating: the set of the lovers of each element in the union of women together with dogs is the intersection of the set of the lovers of all women with the set of the lovers of all dogs. Another example, modernizing notation slightly:

$$l^x = 1 - (1 - l)x, \text{ or } l^x = \overline{\overline{l}x}$$

or, the set of lovers of all women is the complement of the set of individuals who each do not love at least one woman. Herein lies an ambiguity typical of Peirce's 1870 paper, in that the symbol "1" stands for two different things: the first occurrence of "1" refers to the universe U of individuals,

---

[1] Peirce uses lower case italic letters for binary relations and lower case non-italic letters for classes or sets of individuals, although this is not entirely consistent.

whereas the second refers to the Cartesian product U × U, so that the above equation can be written with greater precision (but less elegance) as:

$$l^{\text{x}} = \text{U} - (\text{U} \times \text{U} - l)\text{x}$$

Some other interesting examples to be found in the Logic Notebook are:

$$1^{\text{x}} = 1, \quad 0^{\text{x}} = 0, \quad \text{and} \quad (l \cdot m)^{\text{x}} = l^{\text{x}} \cdot m^{\text{x}}.$$

Again, note that in the first equation, the "1" on the left hand side denotes U × U, whereas on the right hand side it denotes U; also, in the second equation the "0" on the left hand side denotes the empty relation, but on the right hand side it denotes the empty set.

During November and December of 1868, Peirce wrote a set of unpublished notes on various subjects that he had treated in papers he published in 1867. In 'Note 4', regarding the 1867 paper on Boole's calculus, Peirce began by saying:

> The mode proposed for the expression of particular propositions is weak. What is really wanted is something much more fundamental. Another idea has since occurred to me which I have never worked out but which I can here briefly explain. (Peirce, 1984, p. 88)

He was now able to use the exponential function, as explained below. Here is the solution he gave to the problem quoted from De Morgan, above. In these notes to himself, Peirce made a serious mistake in working out his intuition about $0^{\text{x}}$, and this mistake is explained below.

Let $m$ be man, $a$ animal.[2] Then, every man is an animal; or

$$m = m \cdot a$$

or

$$m \cdot 0^a = 0$$

or

$$m + a = a$$

Then,

$$0^a = 0^{m+a} = 0^m \cdot 0^a$$

or

$$0^m = 0^a + 0^m$$

And in the same way, if $h$ denote head,[3]

$$(0^h)^m = (0^h)^a + (0^h)^m$$

---

[2] Because "man" and "animal" are names of classes, m and a should not be italicized.

[3] This is misleading because "head" is the name of the class of heads. As a two-place relation, it should be called, "head of" or "is the head of".

and then

$$0^{\left(0^h\right)^m} = 0^{\left(0^h\right)^a + \left(0^h\right)^m} = 0^{\left(0^h\right)^a} \cdot 0^{\left(0^h\right)^m}$$

That is, any man's head is an animal's head. This result cannot be reached by any ordinary forms of Logic or by Boole's Calculus. (Peirce, 1984, p. 90)

Peirce was attempting to verify that if $m$ is a subset of $a$ then $hm$ is a subset of $ha$, by relying on one of the two ways of expressing the subset relation in Boole's algebra—namely, $m = m \cdot a$. However, he confused the relation $n$ (by which he meant "other than" or "$\neq$") with the empty or zero relation, and then he used this in an exponent notation for the set-theoretic complement (1984, p. 89). Note that $n^x = \bar{x}$:

$$
\begin{aligned}
n^x &= \{i \mid (\forall j)[j \in x \to (i,j) \in n]\} \\
&= \{i \mid (\forall j)[j \in x \to i \neq j]\} \\
&= \{i \mid i \notin x\} \\
&= \bar{x}
\end{aligned}
$$

Let us consider the generalization of this to the case in which the exponent is a relation. By $0^h$ Peirce really meant the relation,

$$
\begin{aligned}
n^h &= \{(i,k) \mid (\forall j)[(j,k) \in h \to (i,j) \in n]\} \\
&= \{(i,k) \mid (\forall j)[(j,k) \subset h \to i \neq j]\} \\
&= \{(i,k) \mid (\forall j)[(j,k) \notin h \lor i \neq j]\} \\
&= \{(i,k) \mid (i,k) \notin h\} \\
&= \bar{h}.
\end{aligned}
$$

Thus the last display in the quotation above,

$$0^{\left(0^h\right)^m} = 0^{\left(0^h\right)^a + \left(0^h\right)^m} = 0^{\left(0^h\right)^a} \cdot 0^{\left(0^h\right)^m}$$

means, using the precise notation explained early in the next section below:

$$h^m = \overline{\bar{h}^m} = \overline{\bar{h}^a + \bar{h}^m} = \overline{\bar{h}^a} \cdot \overline{\bar{h}^m} = h^a \cdot h^m.$$

It has already been mentioned above that in 'Note 4' Peirce used the exponential function to handle the particular syllogism—that is, to express non-emptiness by means of an equation. In 1870 he called the use of this exponential notation, "involution" (1933a, pp. 30–31; 1984, p. 362). In his Logic Notebook, on October 15, 1869, Peirce made the following conjectures about involution (1984, p. 301):

$$l^1 = 1 \ \textit{usually.}$$

$$l^0 = 1 \ \textit{invariably, I think.}$$

In his paper of 1870 he got correct answers to questions of this sort about the exponential function, and by 1873 he showed great confidence in the answers, as we shall see in Section 2.7 below. We examine Peirce's road from conjecture to confidence in Sections 2.6 and 2.7, where we analyze the application of his proof that

$$y^0 = \mathrm{U}$$

to the equation for non-emptiness:

$$0^\mathrm{x} = \left\{ \begin{array}{ll} \mathrm{U} & \text{if } \mathrm{x} = \emptyset \\ \emptyset & \text{if } \mathrm{x} \neq \emptyset \end{array} \right.$$

(1933a, p. 50; 1984, p. 382).

## 2.3   The 1870 relational algebra

Peirce's 1870 paper bears unmistakable marks of haste, insofar as there are missing steps, unacknowledged assumptions, and moreover, inconsistencies in the complicated typographical scheme that Peirce set out to follow but to which he did not always adhere. Naturally, these hindrances to understanding made it unlikely that the paper would be understood—and impossible that its importance and influence would be adequately evaluated—by anyone unfamiliar with the relevant areas of mathematics and mathematical logic. Then there is also Peirce's insistence on non-mathematical examples, which he adopted from De Morgan. These hindrances to understanding Peirce's work are discussed in Section 6.2 below.

Peirce stated at the beginning of his 1870 paper that he intended to combine certain work of Boole and De Morgan, as was quoted at the beginning of the present chapter. He combined the class operations of Boole with the relational operations of De Morgan to get a two-sorted algebra of classes and binary relations, although he actually went beyond Boole's original program of using substitutions in an equational calculus, as we shall discuss in Section 2.7. At the start of his 1870 paper, Peirce separated relations into three categories: absolute terms (unary or monadic relations), binary relations, and higher-order relations (1933a, pp. 33–34; 1984, pp. 365–366). Peirce did not develop a calculus for the higher-order relations in the 1870 paper, and we do not consider them further.

We now proceed to give a Stage One description of Peirce's algebra, making use of set-theoretic concepts:

33

> I propose to use the term "universe" to denote that class of
> individuals *about* which alone the whole discourse is understood
> to run. The universe, therefore, in this sense, as in Mr. De
> Morgan's, is different on different occasions. (Peirce, 1933a, p.
> 35; 1984, p. 366)

We use U to denote the universe, the domain of individuals. As we
mentioned in the previous section, Peirce used the symbol "1" ambiguously
to denote U or U × U. There are two sorts of objects: first, classes or
subsets of the universe U, denoted by lower case Roman letters w, x, y,
z, etc.; and second, relations or subsets of U × U, denoted by lower case
italic letters $w$, $x$, $y$, $z$, etc. In some places, Peirce used upper case letters
for subsets or subrelations; for example, Z for a singleton in z and $X$
for a subrelation of $x$. We will use $\emptyset$ for the empty class and 0 for the
empty relation; Peirce used 0 for both. Set-theoretically, the notions are
identical, but in the algebra they need to be distinguished. The three
Boolean operations of union, intersection, and complement ("+", "·", and
overbar) operate on either sort of objects. These operations are closed on
each of the two sorts of objects, and the ambiguous use of the operation
symbols is context-sensitive. Peirce gave arguments by example for the
standard Boolean facts of commutativity, associativity, distributivity, etc.
(1933a, pp. 28–33, 47–49; 1984, pp. 360–364, 379–381)

Peirce combined these three Boolean operations of union, intersection,
and complement with De Morgan's two operations of relative product and
converse (that is, inverse):

> I shall adopt for the conception of multiplication *the application
> of a relation*, in such a way that, for example, $l$w shall denote
> whatever is lover of a woman. This notation is the same as
> that used by Mr. De Morgan, although he appears not to have
> had multiplication in his mind. (Peirce, 1933a, p. 38; 1984, p.
> 369)

By a "relative", Peirce meant essentially what we mean today by a relation
on a universe U. The relative product is denoted by juxtaposition; for two
relations $s$ and $b$ it can be defined set-theoretically as,

$$sb = \{(i, k) \mid (\exists j)[(j, k) \in b \wedge (i, j) \in s]\};$$

for a relation $s$ and a class w it can be defined as,

$$sw = \{i \mid (\exists j)[j \in w \wedge (i, j) \in s]\}.$$

Note well the appearance of the existential quantifier here, although we
must remind the reader that our description is anachronistic: Peirce did
not have variables for individual elements until 1883, and he did not
have genuine quantifiers until 1885. He gave examples of these operations

using relations between human beings: given that $s$ is "servant of", $b$ is "benefactor of", and w is "women"; then the relation $sb$ is "servant of a benefactor of", the class $sw$ is "servants of women", and the inverse of $s$ is the relation "served by" (1933a, p. 38; 1984, pp. 369–370). The relative product distributes over both the union of classes and the union of relations:

$$s(\mathrm{m} + \mathrm{w}) = sm + sw,$$

$$(l + s)\mathrm{w} = l\mathrm{w} + s\mathrm{w}.$$

The relative product is also associative:

$$(sl)\mathrm{w} = s(l\mathrm{w})$$

(1933a, p. 38; 1984, pp. 369–370).

Peirce also introduced the exponential operation, which he called involution as discussed in Section 2.2 above,

$$s^{\mathrm{w}} \text{ is "servant of each (every) woman",}$$

and which can be defined in modern notation,

$$s^{\mathrm{w}} = \{\mathrm{i} \mid (\forall \mathrm{j})[\mathrm{j} \in \mathrm{w} \to (\mathrm{i}, \mathrm{j}) \in s]\}.$$

He mentioned associativity and two laws of exponents for this involution operation:

$$\left(s^l\right)^{\mathrm{w}} = s^{(l\mathrm{w})}$$

$$s^{\mathrm{m+w}} = sm \cdot sw$$

$$(s \cdot l)^{\mathrm{w}} = s^{\mathrm{w}} \cdot l^{\mathrm{w}}$$

(1933a, pp. 45–46; 1984, pp. 377–78). We need to remark here that Peirce used natural language examples from common experience in connection with these identities, and it is not clear whether he meant merely to exemplify or actually to justify the identities in this way. Recall that De Morgan was quoted in Section 2.1 above as saying, "there are more reasons than one why mathematical examples are not well suited for illustration. The most apposite instances are taken from the relations between human beings..." (De Morgan, 1966, p. 220). This difficulty for a Stage One understanding of Peirce's mathematical meaning will be discussed in Section 2.7 below.

## 2.4   Analysis of part of the 1870 paper

The purpose of Sections 2.4–2.6 is to provide a Stage One analysis and exposition of a passage (in two parts) from Peirce's 1870 paper; the purpose of Section 2.7 is to give a partial Stage Two analysis and commentary on the passage. The particular passage chosen is important because the two proofs in the passage provide one of the best examples for Stage One analysis in the 1870 paper—both for illustration of the method and for demonstration of the difficulties in applying it to Peirce's writings. Moreover, this passage provides the best example in the 1870 paper for a Stage Two consideration of Peirce's lack of today's concepts of set theory: how Peirce tried to proceed without these concepts is both interesting and instructive, and it sheds light on just how difficult these concepts, which we take for granted today, actually are.

The passage comprises Peirce's proofs that, for any relations $x$ and $y$,

$$x\emptyset = \emptyset,$$

and

$$y^{\emptyset} = U$$

(1933a, pp. 50–51; 1984, pp. 382–383). From the modern, set-theoretic view, these two results are rather trivial. Indeed, concerning the first equation, note

$$x\emptyset = \{i \mid (\exists j)[j \in \emptyset \wedge (i,j) \in x]\}.$$

Since $j \in \emptyset$ is false for every j, the entire condition is always false regardless of the value of i; hence $x\emptyset = \emptyset$. Concerning the second equation, note

$$y^{\emptyset} = \{i \mid (\forall j)[j \in \emptyset \rightarrow (i,j) \in y]\}$$
$$= \{i \mid (\forall j)[j \notin \emptyset \vee (i,j) \in y]\}.$$

Since $j \notin \emptyset$ is true for every j, the entire condition is always true regardless of the value of i; hence, it is true for all i that i is related by the relation $y$ to every individual in the empty set $\emptyset$. Thus $y^{\emptyset} = U$, and in particular, $0^{\emptyset} = U$.

As has just been shown, the above two equations are easily proved by use of today's set theory, but Peirce's proofs are quite complex. An analysis of Peirce's actual arguments requires Stage One reconstructions of his concepts; this leads to Stage Two questions about why Peirce felt that he needed such arguments. Nowadays, reasoning about the empty set, the universe, and the definitions of relation and class is familiar to all mathematicians; such reasoning is often learned in association with the quantifiers, relation symbols, and variables used to express that reasoning. Peirce himself did not have quantificational logic until 1883, at the earliest.

## 2.5  Relative product of a relation with the empty class

We now examine Peirce's proof of $x\emptyset = \emptyset$:

> Any relative $x$ may be conceived as a sum of relatives $X$, $X'$, $X''$, etc., such that there is but one individual to which anything is $X$, but one to which anything is $X'$, etc. Thus, if $x$ denote "cause of," $X$, $X'$, $X''$ would denote different kinds of causes, the causes being divided according to the differences of the things they are causes of. Then we have
>
> $$Xy = X(y \,\dagger\, 0) = Xy \,\dagger\, X0,$$
>
> whatever $x$ may be. Hence, since $y$ may be taken so that
>
> $$Xy = 0,$$
>
> we have
> $$X0 = 0;$$
>
> and in a similar way,
>
> $$X'0 = 0, \quad X''0 = 0, \quad X'''0 = 0, \text{ etc.}$$
>
> We have, then,
>
> $$x0 = (X \,\dagger\, X' \,\dagger\, X'' \,\dagger\, X''' \,\dagger\, \text{ etc.})\,0$$
> $$= X0 \,\dagger\, X'0 \,\dagger\, X''0 \,\dagger\, X'''0 \,\dagger\, \text{ etc.}$$
> $$= 0.$$

(Peirce, 1933a, p. 50; 1984, p. 382)[4]

Peirce began the argument by defining the decomposition of a given binary relation with respect to the second place, or in modern notation,

$$X^{(n)} = \{(i, u_n) \,|\, (i, u_n) \in x\},$$

where $\{u_1, u_2, u_3, \ldots\} = U$. (This notation assumes that U is a finite or countable set; see Section 2.7 below.) Hence, using "+" for Peirce's "$\dagger$" and indexing U starting with 1,

$$x = X' + X'' + X''' + \ldots$$

---

[4]NB: from the beginning of Section 2.5 to here, a lower-case italic "$y$" has been used to denote a binary relation. The lower-case non-italic "y" that is introduced below is a one place class or set.

By previous arguments, Peirce showed that, for any class y,

$$y + \emptyset = y$$

(1933a, p. 50; 1984, p. 382), and that for any relation $x$ and any classes y and z,

$$x(y + z) = xy + xz$$

(1933a, p. 47; 1984, p. 379). In particular, for any j,

$$X^{(j)}y = X^{(j)}(y + \emptyset) = X^{(j)}y + X^{(j)}\emptyset. \tag{2.1}$$

Assuming that the universe has at least two elements, we may choose a non-empty class y so that $u_j \in y$ (take y to be any singleton $\{u_k\}$ such that $k \neq j$) and therefore, $X^{(j)}y = \emptyset$, which implies $X^{(j)}\emptyset = \emptyset$ by (2.1). Hence,

$$\begin{aligned}
x\emptyset &= (X' + X'' + X''' + \ldots + X^{(j)} + \ldots)\emptyset \\
&= X'\emptyset + X''\emptyset + X'''\emptyset + \ldots + X^{(j)}\emptyset + \ldots \\
&= \emptyset,
\end{aligned}$$

where j ranges over the indices of elements of U. Peirce did not mention, in the 1870 paper, whether he meant that the union could be taken over an infinite family or that distributivity could be applied to an infinite sum. This apparent departure from Boole's program and the issue of countability are discussed in Section 2.7 below.

## 2.6 Involution of a relation with the empty class

Now we examine Peirce's proof of $y^\emptyset = U$:

> If the relative $x$ be divided in this way into
>
> $$X, X', X'', X''', \text{etc.,}$$
>
> so that $x$ is that which is either $X$ or $X'$ or $X''$ or $X'''$, etc., then non-$x$ is that which is at once non-$X$ and non-$X'$ and non-$X''$, etc.; that is to say,
>
> $$\text{non-}x = \text{non-}X, \text{non-}X', \text{non-}X'', \text{non-}X''', \text{etc.,}$$
>
> where non-$X$ is such that there is something $(Z)$ such that everything is non-$X$ to $Z$; and so with
>
> $$\text{non-}X', \text{non-}X'', \text{etc.}$$

Now, non-$x$ may be any relative whatever. Substitute for it, then, $y$; and for

$$\text{non-}X, \text{ non-}X', \text{ etc., } Y, Y', \text{ etc.}$$

Then we have

$$y = Y, \ Y', \ Y'', \ Y''', \text{ etc.;}$$

and

$$Y'Z' = 1, \ Y''Z'' = 1, \ Y'''Z''' = 1, \text{ etc.,}$$

where $Z'$, $Z''$, $Z'''$ are individual terms which depend for what they denote on $Y'$, $Y''$, $Y'''$. Then we have

$$1 = Y'Z' = Y'^{Z'} = Y'^{(Z'\dagger 0)} = Y'^{Z'}, \ Y'^0 = Y'Z', \ Y'^0$$

or

$$Y'^0 = 1, \ Y''^0 = 1, \ Y'''^0 = 1, \text{ etc.}$$

Then

$$y^0 = (Y', \ Y'', \ Y''', \text{ etc.})^0 = Y'^0, \ Y''^0, \ Y'''^0, \text{ etc.} = 1.$$

(Peirce, 1933a, p. 51; 1984, p. 383)[5]

Peirce began this argument by taking the set-theoretic complement of the relation $x$ and applying one of De Morgan's laws to the decomposition,

$$x = X' + X'' + X''' + \ldots,$$

and hence, using "$\cdot$" for Peirce's "," or set-theoretic intersection,

$$\bar{x} = \overline{(X' + X'' + X''' + \ldots)} = \overline{X'} \cdot \overline{X''} \cdot \overline{X'''} \ldots$$

Of course, this requires the infinite application of De Morgan's law if the universe is infinite. So long as there are at least two elements in the universe, then for a given j there is some k such that $k \neq j$. Since $X^{(j)}$ contains none of the pairs $\{(u_j, u_k) \mid k \neq j\}$, the complement of $X^{(j)}$ contains all such pairs. Now, for convenience, Peirce renamed $\bar{x}$ as $y$ and $\overline{X^{(j)}}$ as $Y^{(j)}$, so that,

$$y = Y' \cdot Y'' \cdot Y''' \cdot \ldots \tag{2.2}$$

Each $Y^{(j)}$ has an associated singleton $Z^{(j)}$ which is some $\{u_k\}$ such that $k \neq j$. Peirce called these singletons, "individual terms", and we discuss this in the next section. Therefore, for each j,

$$U = Y^{(j)} Z^{(j)} = (Y^{(j)})^{Z^{(j)}}$$

---

[5]Typographical corrections: $Z$, a singleton class, should not have been italicized in the original; the series $Y$, $Y'$, $Y''$, ... became, without explanation, $Y'$, $Y''$, $Y'''$, ..., but we shall use the latter consistently.

because, since $Z^{(j)}$ is a singleton, the set of i such that i is related to some element of $Z^{(j)}$ is the same as the set of i such that i is related to every element of $Z^{(j)}$. Continuing,

$$(Y^{(j)})^{Z^{(j)}} = (Y^{(j)})^{(Z^{(j)}+\emptyset)}$$
$$= (Y^{(j)})^{Z^{(j)}} \cdot (Y^{(j)})^{\emptyset}$$
$$= Y^{(j)} Z^{(j)} \cdot (Y^{(j)})^{\emptyset}.$$

Therefore, since

$$U = Y^{(j)} Z^{(j)} \cdot (Y^{(j)})^{\emptyset},$$

we must have

$$U = (Y^{(j)})^{\emptyset}. \tag{2.3}$$

Finally, from (2.2), (2.3), and the formula[6]

$$(x \cdot y)^z = x^z \cdot y^z$$

extended to arbitrarily many factors, we have,

$$y^{\emptyset} = (Y' \cdot Y'' \cdot Y''' \cdot \ldots)^{\emptyset} = (Y')^{\emptyset} \cdot (Y'')^{\emptyset} \cdot (Y''')^{\emptyset} \cdots = U.$$

If the universe is infinite, the products have infinitely many factors.

## 2.7 Comments on Peirce's proofs

At the end of Section 2.2, we mentioned that the problem of $0^{\emptyset}$ was important for Peirce because he wanted to use this to express the non-emptiness of a class—that is, to solve the problem of expressing "there exists", as he wrote in the 1870 paper:

> That which first led me to seek for the present extension of Boole's logical notation was the consideration that as he left his algebra, neither hypothetical propositions nor particular propositions could be properly expressed. (Peirce, 1933a, p. 90; 1984, p. 421)

> What is wanted, in order to express hypotheticals and particulars analytically, is a relative term which shall denote "case of the existence of —," or "what exists only if there is any —"; or else "case of the non-existence of —," or "what exists only if there is not —." When Boole's algebra is extended to relative terms, it is easy to see what these particular relatives must be. ...Now, $0^x$ is such a function, vanishing when $x$ does not, and not vanishing when $x$ does.[7] (1933a, p. 91; 1984, p. 423)

---

[6] See 1933a, p. 48; 1984, p. 380

[7] Peirce is using "0" to mean the empty binary relation, and he is dropping the convention that lower case italic letters stand for relations. Here we shall follow the earlier convention, and x shall not be italic.

As we saw in Section 2.1, Boole had attempted to express "the existence of —" with the term $v$, which signified an indefinite non-empty class (1847, p. 21). Peirce applied the results analyzed in the previous two sections to solve the particular syllogism in the 1870 paper, where he proved from the premisses, 'Every horse is black', 'Every horse is an animal', and 'There are some horses', the conclusion that 'Some animals are black' (1933a, p. 93; 1984, pp. 424–425). By 1872[8], Peirce was confidently using two equations to express non-emptiness:

> But 1 and 0 have sometimes to be interpreted as relative terms. Now, it can be proved by the principles of the logic of relatives that so considered $0^x = 0$, unless x = 0, when $0^0 = 1$; and that 1x = 1, unless x = 0, when 10 = 0. It follows that $0^x$ is such a logical function of x that it signifies "the case of the non-existence of," while 1x is such a logical function of x that it signifies "the case of the existence of." (Peirce, 1976c, p. 640; 1986a, p. 115)

Thus, Peirce used either of these two equations in the relational algebra to express non-emptiness:[9]

$$0^x = \left\{ \begin{array}{ll} U & \text{if } x = \emptyset \\ \emptyset & x \neq \emptyset \end{array} \right.$$

$$(U \times U)x = \left\{ \begin{array}{ll} U & \text{if } x = \emptyset \\ \emptyset & x \neq \emptyset \end{array} \right.$$

We now return to the Stage Two question posed in Section 2.4: Why did Peirce use such complex reasoning to prove what seems so trivial to us today? The contemporary proofs given in Section 2.4 are easy exercises in simple set theory. In 1870, Peirce lacked the distinction of an object from its singleton, the explicit "element of" relation, and the confident use of the empty set as an object; he definitely did not have the quantifiers, nor did he have a facile use of the propositional calculus—especially material implication. Instead of quantifiers and propositional calculus, we find algebraic manipulations that seem less direct and less clear. Thus we have

---

[8]Peirce wrote the paper, 'On the Theory of Errors of Observation', in late 1872; the publication date is given as 1870; and the paper was actually printed in 1873 (1986a, p. 515). His enthusiasm for these equations was so great that he included the above passage at the beginning of a paper, the object of which was, "to give a general account of the theory of errors of observations, with the design of showing what the limitations to the applicability of the method of least squares are, and what course is to be pursued when that method fails." (1976c, p. 639; 1986a, p. 114)

[9]Peirce's ambiguous use of "0" and "1" to be classes or binary relations, depending on context, have been clarified in the following equations as explained in Section 2.3 above. Thus, U is the class meaning of 1, and U × U is the relation meaning of 1; $\emptyset$ is the class meaning of 0, and the null or empty relation is the relation meaning of 0.

both a negative and a positive question: first, why is set theory missing? and second, why do we find algebra instead?

What Peirce did have was his explicitly stated plan to combine De Morgan's work on relations with Boole's work—especially Boole's program for an equational calculus of deductive reasoning limited only to equations and substitutions in the equations—as discussed earlier in this chapter.

In the proof of $x\emptyset = \emptyset$, analyzed in Section 2.5 above, Peirce began by decomposing the binary relation $x$ with respect to its second place, but his explanation of the decomposition is not at all clear to the modern reader. In the quotation at the beginning of Section 2.5, his use of the relation "cause of" to indicate the first and second places in the set of ordered pairs is confusing:

> Thus, if $x$ denote "cause of," $X$, $X'$, $X''$ would denote different kinds of causes, the causes being divided according to the differences of the things they are causes of.[10] (Peirce, 1933a, p. 50; 1984, p. 382)

Nowadays, we are so used to set theory that no example would even be necessary. Perhaps Peirce thought that "cause of" would more strongly suggest a first and second place, or perhaps he could not find a human relation to serve as a satisfactory example. In any case, Peirce's decomposition as described by today's notions is:

$$x = \cup_j\{(i, u_j) \mid (i, u_j) \in x\} = \cup_j X^{(j)},$$

where $U = \{u_1, u_2, u_3, \ldots\}$. He used upper case $X$ to indicate that the $X^{(j)}$ were obtained from a given relation $x$ as disjoint subrelations of $x$. Then he claimed that for each $X^{(j)}$ it is possible to find a set y such that $u_j \notin y$. But note that he was not willing to take $y = \emptyset$; otherwise, no proof by decomposition would even have been necessary. Using the identity $y = y + \emptyset$, he substituted $y + \emptyset$ for y in order to force $x\emptyset = \emptyset$. He used lower case y because y could be any subset of $U - \{u_j\}$. Thus, he avoided using $\emptyset$ as a legitimate class; instead, he used it as a formal element in an equational calculus, thereby following Boole's program by performing substitutions according to established identities.

In the proof of $y^\emptyset = U$ that was analyzed in Section 2.6, Peirce applied De Morgan's law to the decomposition of $x$ into subrelations $X^{(j)}$. Rather unfortunately, in that proof and in the early part of the proof in Section 2.5 he chose to use lower-case italic "$y$" as the complement of $x$, confusing the use of "y" as a class in the latter part of the proof in Section 2.5 and elsewhere with the use of "y" as a relation. The decomposition into a union thus became a decomposition into an intersection:

$$\bar{x} = y = \cap_j\{(i, u_j) \mid (i, u_j) \notin x\} = \cap_j Y^{(j)} = \cap_j \overline{X^{(j)}}.$$

---

[10]Now the lower case italic $x$ stands for a relation.

However, in this argument Peirce needed a singleton $Z^{(j)} \neq \{u_j\}$ for each $\overline{X^{(j)}}$ in order to apply substitution according to the identity

$$Y^{(j)} Z^{(j)} = (Y^{(j)})^{Z^{(j)}},$$

which holds only if $Z^{(j)}$ is a singleton. But note that he used upper case Z by analogy with upper case $X$ in the decomposition of the relation $x$. He was reluctant to treat a singleton as a legitimate class: otherwise, he would have used a lower case z as he used a lower case y in the previous proof.

Finally, if $x$ in the first proof is legitimate (that is, non-empty but possibly U × U), then $y$ in the second proof is a strict subset of U × U but nonetheless may be empty. Thus, the proof that $y^0 = $ U yields $0^0 = $ U as a special case, and together with $0^x = \emptyset$ for x $\neq \emptyset$, he had proved the equation for non-emptiness,[11]

$$0^x = \begin{cases} \text{U} & \text{if x} = \emptyset \\ \emptyset & \text{if x} \neq \emptyset \end{cases}$$

which was so important to him.

When Peirce introduced into the arguments analyzed above such expressions as,

$$X' + X'' + X''' + \ldots,$$

he departed from Boole's original program. Peirce does not mention in the 1870 paper whether he meant that unions could be taken over infinite families or that distributivity could be applied to an infinite sum. If he did admit the infinite case, this was an obvious departure from Boole's program of a purely equational calculus. But even if he only meant to restrict the use of the ellipsis (given as "etc." in the original) to signify finite, indefinitely large expressions, this would still have been a departure from Boole's program to the extent that mathematical induction in the metalanguage would be necessary to provide a meaning for such expressions as,

$$X' + X'' + X''' + \ldots + X^{(n)}.$$

Of course, to suggest that Peirce could have viewed the matter in such a way would be preposterous anachronism. Furthermore, the equation,

$$\bar{x} = \overline{(X' + X'' + X''' + \ldots)} = \overline{X'} \cdot \overline{X''} \cdot \overline{X'''} \cdot \ldots,$$

discussed in Section 2.6, required the application of De Morgan's law to an infinite or indefinitely large expression, which was yet another departure from Boole's program.

What we definitely do have in Peirce's approach is his efforts to prove his results without the set-theoretic techniques we possess today. It is

---

[11] A non-italic x is needed to here to be compared with the empty class.

anachronistic to question his competence because he lacked these techniques, notwithstanding their present place in the undergraduate or beginning graduate curriculum. Indeed, we should rather be led to wonder just how elementary or trivial or obvious these techniques actually are.

In 1870, Cantor had not yet published his first paper on set theory. Peirce did not mention Cantor in print until 1889 (1935, p. 114), and in or about 1905 he implied that he did not know of Cantor's work before 1883 (1934, pp. 351n, 368n2). He mentioned Dedekind as early as 1897 in a review of Schröder's work (1933a, p. 333).

# Chapter 3

# Peirce's Quantificational Logic

Peirce was led to the discovery of the quantifiers by representing his relation algebra as an algebra of matrices. The process began in his paper of 1870, where he considered the effect of the relative product on singleton relations and got an associative multiplication on ordered pairs of individuals. Extending this multiplication to linear combinations of ordered pairs, he got an associative algebra. In *Brief Description of the Algebra of Relatives*, a privately printed monograph of 1882, he expressed the product of two elements of this algebra in terms of their coefficients with respect to basis elements (1933a, pp. 180–186; 1986b, pp. 328–333); thus he got what today is called the formula for matrix multiplication, although he apparently did not know about matrices at the time. In 'Note B: The Logic of Relatives', one of his contributions to the 1883 collection, *Studies in Logic, By Members of the Johns Hopkins University*, he allowed the entries in the matrices to be truth values, thereby getting a representation of the algebra of binary relatives (1933a, pp. 195–209; 1986b, pp. 453–466). He observed that in the matrix representation of the relative product the notions of 'for some' (or existence) and 'for all' are expressed as the non-vanishing of a certain sum or product of truth values, respectively. Whereas in his 1883 paper the symbols $\sum$ and $\prod$ denoted the sum and product of truth values, in his 1885 paper, 'On the Algebra of Logic: A Contribution to the Philosophy of Notation', Peirce reinterpreted the symbols $\sum$ and $\prod$ to stand for the actual notions of 'for some' and 'for all' (1933a, pp. 210–238, esp. 228). To an extent hardly realized, Peirce arrived at his conception of the quantifiers through considerations of abstract algebra. In this way, he provided the formalism and semantics of the system of quantificational logic upon which the 1915 paper of Löwenheim and the 1920 and 1923 papers of Skolem were based.

The details of Peirce's part in this story concern us in the present chapter; the work of Löwenheim and Skolem will be discussed in Chapter 6.

## 3.1  More algebra of binary relations

Near the end of the 1870 paper Peirce considered the effect of the relative product on what he then called elementary relatives and later, individual relatives (1933a, pp. 75–81; 1984, pp. 408–414). By a "relative", he meant what we mean today by a relation on a universe U; by "individual relatives", he meant the singleton sets of ordered pairs denoted $(u_i:u_j)$, where $u_i$ and $u_j$ are elements of the universe or domain of individuals. Although our conception is anachronistic to the extent that it is more purely extensional than Peirce's was in 1870, it is close enough to what Peirce meant, and it provides a coherent interpretation—especially with respect to the hindsight provided by our heritage of ideas developed from Peirce's work.

Since the relative product is associative, the system of individual relatives taken over a universe $U$ is an associative system under the following law of multiplication:

$$(u_i : u_j) * (u_k : u_l) = \begin{cases} (u_i : u_l) & \text{if } j = k \\ 0 & \text{if } j \neq k \end{cases}$$

where 0 stands for the empty relation (1933a, p. 77; 1984, p. 410). There are $n^2$ ordered pairs over a finite universe of n individuals. Hence, including 0, our associative system has $n^2 + 1$ elements, and the relative multiplication table has $n^2+1$ rows and columns—the 0 row and column containing only zeros. Peirce's simplest example was based on a 2 element universe $U = \{u, v\}$ wherein u is a teacher and v is a pupil. The individual relatives are: $(u : u) = c$, "colleague of"; $(u : v) = t$, "teacher of"; $(v : u) = p$, "pupil of"; and $(v : v) = s$, "schoolmate of" (1933a, pp. 76–78; 1984, pp. 409–411). There are $2^2$ individual relatives and, including the 0 relative, $2^2 + 1$ rows and columns in the complete table. Thus, each universe of n individuals leads to an associative multiplication table for the $n^2+1$ individual relatives. Peirce then took formal linear combinations of these $n^2 + 1$ individual relatives, with scalars taken from the real or complex numbers, and thereby obtained an associative algebra $P_n$ of dimension $n^2$.

In subsequent years, $P_n$ was for Peirce not only an algebra, but indeed it was an algebra of linear transformations. Already in an 1873 manuscript for part of a proposed book on logic, he regarded the $(u_i : u_j)$ as mappings of $U$ into $U$:

$$(u_i : u_j) \, u_k = \begin{cases} u_i & \text{if } u_j = u_k \\ \text{absurd} & \text{otherwise} \end{cases}$$

(1986a, pp. 93–95). But more correctly, he should have said,

$$(u_i : u_j) u_k = \begin{cases} u_i & \text{if } u_j = u_k \\ \emptyset & \text{otherwise} \end{cases}$$

As we remarked in Section 2.7, Peirce had such difficulties both because he lacked a clear distinction between element and singleton set and because he was uneasy about the use of the empty class. He was now in a position to take formal linear combinations of elements of $U$ and to extend the $(u_i : u_j)$ to linear transformations on these formal linear combinations. Thus $P_n$ is not only an algebra, but an algebra of linear transformations.

## 3.2   1881 representation theorem and matrices

In 'On the Application of Logical Analysis to Multiple Algebra', a brief paper of 1875, Peirce made the first mention of his representation theorem; he said at the beginning of the paper that he would consider the algebra as acting on itself by left multiplication (1933a, pp. 99–101; 1986a, pp. 177–179). In 1881, Peirce edited his father's privately printed monograph, *Linear Associative Algebra* (B. Peirce, 1870), for publication in the *American Journal of Mathematics* (B. Peirce, 1881). In the third of four appended notes, he proved this important result: "…any associative algebra can be put into relative form, *i.e.* …every such algebra may be represented by a matrix." (1933a, p. 173; 1986b, p. 321) Peirce's proposition may be restated in contemporary mathematical terms as,

> **Theorem 1.**  Every associative algebra of dimension n is isomorphic to a subalgebra of $P_{n+1}$.

The ideas behind his proof of this theorem are as follows: The $n^2 + 1$ individual relatives, including the 0 relative, form an associative system or semigroup $S_n$. By taking formal linear combinations of elements of this semigroup, with coefficients from a given field—for Peirce, the real or complex numbers—we get the semigroup algebra $A_n$ in which there are three operations: scalar multiplication, vector addition, and vector (that is, matrix) multiplication. This is the technique of the group algebra, which actually requires only that the underlying system be a semigroup. Van der Waerden says in his *History of Algebra*:

> In the same paper of 1854, in which Cayley introduced the notion of an abstract group (Phil. Mag. 7, p. 40–47), he also introduced what we today call the "Group Algebra" of a finite group $G$. The basis elements of this algebra are just the group elements $g_1, \ldots, g_n$. In the multiplication rule
>
> $$g_j g_k = \sum g_i a_{ijk}$$
>
> the coefficients are
>
> $$a_{ijk} = \begin{cases} 1 & \text{if } g_j g_k = g_i \\ 0 & \text{otherwise} \end{cases}$$

Every representation of the group $G$ by linear transformations can be extended to a representation of the group algebra. Conversely, every representative of the group algebra yields a representation of the group. Therefore the study of the structure of the group algebra is of primary importance in the theory of group representations. (van der Waerden, 1985, p. 190)

Peirce was a pioneer in this technique, which is important today in representation theory. Bourbaki, in *Éléments d'histoire des mathématiques*, gives Peirce credit for recognizing that the square matrices form an algebra, for giving the regular representation of an algebra as acting on itself, and for defining the group algebra (1960, p. 121)[1]. See also *Theory of Group Representations* (Naimark and Stern, 1982, pp. 87–92).

But $A_n$ has dimension $n^2 + 1$, so we form a quotient algebra that lowers the dimension to $n^2$, thereby obtaining the desired $P_n$. Namely, let B consist of all scalar multiples of the 0 element of $S_n$. Then B is an ideal, and $P_n = A_n / B$.

As stated above, Peirce saw $P_n$ as an algebra of linear transformations. Given $U = \{u_1, \ldots u_n\}$ and $U* = U \cup \{0\}$ , then $U*$ has $n + 1$ elements and each $(u_i : u_j)$ is a mapping $U* \to U*$. Let $V_{n+1}$ be the vector space consisting of all formal linear combinations of elements of $U*$. Then each $(u_i : u_j)$ can be extended in a unique way to a linear transformation on $V_{n+1}$.

Now let $H$ be the subspace of $V_{n+1}$ consisting of all scalar multiples of the special element $\{0\} \in U*$. Let $E_n$ denote the quotient space $V_{n+1}/H$ considered as a vector space. As transformations, all the $(u_i : u_j)$ leave H invariant, and thus they act as linear transformations on $E_n$. Hence, $P_n$ is the set of all linear combinations of these transformations $(u_i : u_j)$.

**Proof of Theorem 1.** Let $L$ be an associative algebra of dimension n. We proceed in two steps:

*Step one.* Observe that each element of $L$ acts on $L$ as a linear transformation by left multiplication. This represents $L$ as an algebra of linear transformations. But this representation may not be faithful, which is to say that some element of $L$ may be represented by the zero transformation. The representation is faithful, however, if $L$ has a multiplicative identity. Whether or not $L$ already has a multiplicative identity, we can adjoin a multiplicative identity to $L$ by a technique that is standard today, thereby getting an algebra $L*$. Using $F$ for the scalar field, $L* = F \times L$,

---

[1] It is widely believed that André Weil wrote most of the historical material contained in the pseudonymously and collectively written volumes attributed to the *nom de plume*, N. Bourbaki. This historical material was collected in the single volume cited.

$$(f_1, l_1) + (f_2, l_2) = (f_1 + f_2, l_1 + l_2),$$
$$f_2(f_1, l_1) = (f_2 f_1, f_2 l_1),$$
$$(f_1, l_1) \cdot (f_2, l_2) = (f_1 f_2, f_1 l_2 + f_2 l_1 + l_1 l_2),$$

and $(1, 0)$ is the identity in $L*$. Now $L$ acts faithfully on $L*$ as an algebra of linear transformations.

*Step two.* First, observe that every algebra of linear transformations of $E_n$ is a subalgebra of $P_n$, because a linear transformation $f : E_n \to E_n$ is determined by its behavior on the basis elements

$$f(u_i) = \sum_j \lambda_{ji} u_j$$

hence,

$$f = \sum_{jk} \lambda_{jk}(u_j : u_k).$$

Now let the algebra $L$ have the basis $\{a_1, \ldots, a_n\}$. In his 1881 paper, Peirce associated a linear transformation of the form

$$(a_i : a_0) + \sum_{jk} \lambda_{jk}(a_j : a_k)$$

to each basis element $a_i$ (1933a, pp. 171–172; 1986b, pp. 319–320). By hindsight, the addition of the term $(a_i : a_0)$ can be understood as corresponding to the extension of to $L*$, as in Step one above.

Q.E.D.

Peirce realized by the spring of 1882 that his algebra $P_n$ is mathematically identical to the algebra of all n × n matrices over the given field. Indeed, writing $(i : j)$ for $(u_i : u_j)$, direct calculation yields the following:

$$\left(\sum_{ij} \alpha_{ij}(i : j)\right)\left(\sum_{kl} \beta_{kl}(k : l)\right) = \sum_{il} \gamma_{il}(i : l)$$

where

$$\gamma_{il} = \sum_x \alpha_{ix} \beta_{xl} \tag{3.1}$$

The last formula (3.1) is the formula of matrix multiplication. Since Peirce used the word "matrix" in his own statement of the theorem, he must have recognized the connection between his algebra $P_n$ and matrices before the 1881 appendix to his father's paper was actually published in 1882 (Fisch,

1986b, pp. 58–59); while the 1881 appendix was still in press, Peirce was able to add to his statement of the theorem (quoted at the start of the present section) a parenthetical reference to his 1882 paper on the algebra of relatives:

> Thus, what has been proved is that any associative algebra can be put into relative form, i.e. (see my *brochure* entitled *Brief Description of the Algebra of Relatives*) that every such algebra may be represented by a matrix. (Peirce, 1933a, p. 173; 1986b, p. 321)

In his 1882 *Brief Description*, the scalars consisted of "numbers, which may be permitted to be imaginary or restricted to being real or positive, or to being roots of any given equation, algebraic or transcendental." (1933a, p. 180; 1886b, p. 328) Peirce stated in a footnote to this, "I have usually restricted the coëfficients to one or other of two values; but the more general view was distinctly recognized in my paper of 1870." (1933a, p. 180n1; 1986b, p. 328n1)

Peirce's hasty private printing in early 1882 of his *Brief Description* and his two controversies with Sylvester, one over the addition of a credit to Peirce in a paper of Sylvester's and the other over the priority for the discovery of nonions, indicate to us that Peirce was very concerned about how his work would be received in light of Cayley's much earlier work on matrices (1933a, p. 186; 1986b, p. 333). In a manuscript quoted by Max Fisch, Peirce recounted the story of how he showed the *Brief Description* to Sylvester:

> When it was done and I was correcting the last proof, it suddenly occurred to me that it was after all nothing but Cayley's theory of matrices which appeared when I was a boy. However, I took a copy of it to the great algebraist Sylvester. He read it, and said very disdainfully—Why, it is nothing but my umbral notation. I felt squelched and never sent out the copies. But I was a little comforted later by finding that what Sylvester called "my umbral notation" had first been published in 1693 by another man of some talent named Godfrey William Leibniz... (Fisch, 1986b, p. 58)

Thus, although Peirce's 1882 *Brief Description* was printed and used in his courses at Johns Hopkins, it was not effectively published (Fisch, 1986b, pp. 47, 58, 76n56). Evidence of Peirce's resentment over this affair can be seen in an unpublished manuscript of 1902 (1933b, p. 246). For thorough and interesting accounts of these events, see (Fisch, 1986b, pp. 54–62) and (Houser, 1986, pp. xxxvi–lxx, esp. lii–lix).

# 3.3   1883 matrix representation

When Peirce began teaching at Johns Hopkins in 1879, he organized a Metaphysical Club to discuss logic.[2] The meetings usually included talks by students and faculty. During this time, Peirce and his students, especially Christine Ladd (later Ladd-Franklin) and Oscar H. Mitchell, apparently began to work on the problem of systematically incorporating the notions of 'for all' and 'for some' into a theory of logic. Some of their work appears in the 1883 volume edited by Peirce, *Studies in Logic by Members of the Johns Hopkins University*. Both Ladd and Mitchell expressed the notions of quantification in their papers, 'On the Algebra of Logic' (Ladd, 1883) and 'On a New Algebra of Logic' (Mitchell, 1883); but they did not use symbols for quantification which bind variables for individuals. The question of what Mitchell actually did is discussed in Section 3.4.1 below. Peirce was probably stimulated by the ideas he discussed with his students, and perhaps especially by those in Mitchell's paper, to develop his own solution to the problem in 'Note B: The Logic of Relatives', an appendix to *Studies in Logic* (1933a, pp. 195–209; 1986b, pp. 453–466).

In 'Note B', Peirce used truth values instead of field elements in the matrices, giving what is called today the graph of a relation. These truth values, represented by 0 and 1, are added and multiplied in the usual way, except that 1+1=1 (1933a, p. 196; 1986b, pp. 454–455). In formal linear combinations, the operation of addition is actually taken as the addition in the scalar field. Since any relative is the union of its constituent individual relatives, these new linear combinations (with truth value coefficients 0 or 1) give precisely the unions of individual relatives; and thus the operation of addition in this context is actually set-theoretic union. Furthermore, the matrix corresponding to the relative product of two relations $l$ and $b$ is precisely the matrix product of the individual matrices corresponding to $l$ and to $b$.

The connection between the matrix representation and the ideas of quantification is expressed thusly in 'Note B': The relative product,

lb for lover of a benefactor,

is "defined" by means of the equation,

$$(lb)_{ij} = \sum_{x} (l)_{ix}(b)_{xj}$$

(1933a, p. 197; 1986b, p. 455). This is just the formula (3.1) for matrix multiplication which we arrived at in the previous section. Peirce stated that the relative product $lb$ "is called a particular combination, because it implies the *existence* of something *loved by* its relate and a *benefactor of*

---

[2]Peirce probably named it after the Cambridge Metaphysical Club at which he first discussed his concept of pragmatism during 1871–75.

its correlate." Similarly, the relative sum involves a product and the idea of universality:

$$(lb)_{ij} = \prod_x \{(l)_{ix} + (b)_{xj}\}, \tag{3.2}$$

insofar as everything is loved by its relate i or is a benefactor of its correlate j (1933a, p. 197; 1986b, p. 455). Later in the 1883 paper, he returned to these sums and products of numerical coefficients:

> Any proposition whatever is equivalent to saying that some complexus of aggregates and products of such numerical coefficients is greater than zero. Thus,
>
> $$\sum_i \sum_j l_{ij} > 0$$
>
> means that something is a lover of something; and
>
> $$\prod_i \sum_j l_{ij} > 0$$
>
> means that everything is a lover of something. (Peirce, 1933a, pp. 206–207; 1986b, p. 464)

Then he dropped the equational form of the expression:

> We shall, however, naturally omit, in writing the inequalities, the >0 which terminates them all; and the above two propositions will appear as
>
> $$\sum_i \sum_j l_{ij} \text{ and } \prod_i \sum_j l_{ij}.$$

(Peirce, 1933a, p. 207; 1986b, p. 464)

And then followed further examples using this notation (1933a, pp. 207–209; 1986b, pp. 464–466).

Thus Peirce expressed the notion of existence or 'for some' as being equivalent to the non-vanishing of a sum, and he used the subscripts as variables for individuals of the universe. Similarly, the notion of universality or 'for all' was equivalent to the non-vanishing of a product. He represented 'for some' and 'for all' by means of mathematical concepts, but from his discussion of examples, we see that he was well on the way to making the crucial steps of introducing the symbols $\sum_i$ and $\prod_i$ into the formalism and reinterpreting them as *logical* operators or quantifiers—steps made in the 1885 paper discussed in Section 3.5 below. Presumably, the universe is allowed to be infinite in this paper. In the case of a countable universe there is no problem with the meaning of the relative product equation

(3.1), because an infinite series of truth values always converges; similarly, in the relative sum equation an infinite product of truth values always converges. What these formulas might mean in the case of an uncountable universe is another question.

# 3.4   Two historical questions

Here it is appropriate to address two historical questions: Did O.H. Mitchell actually invent the quantifiers, as Peirce stated in a number of places? and, Did Peirce know about Frege's *Begriffsschrift*, published in 1879? Geraldine Brady discusses these two questions (Brady, 1990, pp. 18–26), giving a more thorough account of Mitchell's ideas than is necessary for our purposes. Two of the major, most comprehensive histories of logic state correctly that Mitchell did not have true quantifiers, although they both surmise incorrectly that Peirce had not heard of Frege by 1883 (Bochenski, 1970, pp. 347–348; Kneale and Kneale, 1962, pp. 431–432).

## 3.4.1   Did Mitchell discover the quantifiers?

In a number of published statements, Peirce attributed the discovery of quantifiers to Mitchell. For example, early in his 1885 paper he said,

> The introduction of indices into the algebra of logic is the greatest merit of Mr. Mitchell's system. He writes $F_1$ to mean that the proposition $F$ is true of every object in the universe, and $F_u$ to mean that the same is true of some object. This distinction can only be made in some such way as this. Indices are also required to show in what manner other signs are connected together. (Peirce, 1933a, p. 212)

and immediately preceding his own introduction of quantifiers in the 1885 paper he said,

> We now come to the distinction of *some* and *all*, a distinction which is precisely on a par with that between truth and falsehood; that is, it is descriptive.

> All attempts to introduce this distinction into the Boolian algebra were more or less complete failures until Mr. Mitchell showed how it was to be effected. His method really consists in making the whole expression of the proposition consist of two parts, a pure Boolian expression referring to an individual and a Quantifying part saying what individual this is.

> ...Mr. Mitchell has also a very interesting and instructive extension of his notation for *some* and *all*, to a two-dimensional

> universe, that is, to the logic of relatives. (Peirce, 1933a, pp. 227–228)

These remarks are very misleading, and some writers on logic have been misled by them—for example, see (Putnam, 1982, pp. 294, 297, 298). Although Mitchell had subscripts, they did not stand for individual variables; thus they were not "indices" in Peirce's sense. Furthermore, in Mitchell's paper there are not two symbols, one for the universal quantifier and one for the existential; instead, there is a single subscript symbol (which subscript may be a string) for the act of quantification, the subscript indicating both whether the quantification is universal or existential for each place and what the order of alternation of quantifiers is. In the next section we give an example from one of Peirce's 1880 manuscripts wherein he almost had his 1883 notation.

Using "$F$" for a Boolean combination of monadic predicates, Mitchell's two forms and their modern equivalents are (1883, p. 74):

$$F_1 = (\forall i)F(i),$$

$$F_u = (\exists i)F(i)$$

Using "$F$" for a Boolean combination of binary predicates, his six forms for quantification over binary relations are (*ibid*, p. 87):

$$F_{11} = (\forall i)(\forall j)F(i,j),$$
$$F_{1v} = (\forall i)(\exists j)F(i,j),$$
$$F_{u1} = (\lor j)(\exists i)F(i,j),$$
$$F_{1v'} = (\exists j)(\forall i)F(i,j),$$
$$F_{u'1} = (\exists i)(\forall j)F(i,j),$$
$$F_{uv} = (\exists i)(\exists j)F(i,j)$$

Mitchell considered the two places to represent two different universes or "dimensions", as he called them, giving a two-sorted system. Mitchell acknowledged in a footnote that Peirce pointed out to him how the AE alternation is distinct from the EA; hence, Mitchell had to add the prime marks to get the EA forms (1883, p. 88n). He correctly stated that there are twenty-six quantifier forms for ternary relations, but he gave no indication of how to express the different alternations of quantifiers by means of primes or otherwise (1883, p. 95). This approach would be extremely cumbersome for relations with more than two arguments, and even in the case of binary relations, it does not satisfy Peirce's desire for a convenient calculus.

As Peirce stated in the quotation at the beginning of this section, both he and Mitchell were concerned with expressing propositions in prenex form ("a Quantifying part" followed by "a pure Boolian expression"); later in his paper, Mitchell allowed the propositions with suffix subscripts to be nested. For example,

$$\{(ab)_v + (\bar{c} + \bar{d}_V\}_U$$

(1883, p. 93), means in today's notation,

$$(\forall i)[(\exists j)\{a(i,j) \wedge b(i,j)\} \vee (\forall j)\{\neg c(i,j) \vee \neg d(i,j)\}]$$

The suffix subscripts are operators expressing quantification, upper case letters for universal and lower case for existential; in effect, this approach is similar to that of polyadic or cylindrical algebra (Brady, 1990, p. 24). Thus Mitchell worked on the problem of quantification and used subscript symbols to express quantification, but he did not have the quantifiers as symbols that bind individual variables.

### 3.4.2  Did Peirce know of Frege?

It is not possible to settle this question using the available sources. The facts are: First, Peirce edited the 1883 volume, *Studies in Logic*, which included his, Ladd's, and Mitchell's papers. Second, the bibliography to Ladd's paper included the following entry:

> XVII. Gottlob Frege: Begriffsschrift, eine der arithmetischen nachgebildete Formelsprache des reinen Denkens. Halle, 1879. Reviewed by Ernst Schröder in *Zeitschrift für Mathematik und Physik*, 1880. (Ladd, 1883, pp. 70–71)

Third, Schröder described Frege's expression of quantification in the review, although the account is misleading (Schröder, 1972, pp. 229–230; Brady, 1990, pp. 25–26). And fourth, Peirce owned a copy of Schröder's review of Frege, which was marked in Peirce's hand, 'Formal Logic' (Fisch, 1986b, p. 259n8). Thus, Peirce knew of Frege by 1883, and (presumably by 1883) he had available to him a somewhat faulty account of Frege's quantification.

Nevertheless, as we have seen in Chapter 2 and the present chapter, Peirce had been trying to express particular propositions and existential quantification since 1865. He succeeded in doing so in his 1870 paper, but he continued to try to find a better calculus, which efforts resulted in the expression given in his 1883 paper. Furthermore, as early as 1880 in his manuscript 'Logic Notebook', he almost had the 1883 expression of quantifier and index:

>...let

$$x = \mathbf{f}$$

signify that whatever object can be chosen is not $x$

$$x = \mathbf{v}$$

that whatever object can be chosen is $x$...

$$(x - \mathbf{f})(y - \mathbf{v}) = 0$$

means each object chosen is either not-x or is y. (Peirce, 1986b, p. 216)

Here, Peirce used "$\mathbf{f}$" and "$\mathbf{v}$" as truth values:

I have myself proposed (Logic of Rel p 7) to write $x > y$ to mean $x$'s comprise some objects besides $y$'s but properly this can only mean $x = \mathbf{v}$
$y = \mathbf{f}$ (if $\mathbf{v} > \mathbf{f}$)... But if $x_i$ denotes what $x$ becomes for the $i^{\text{th}}$ object, since

$$(x - \mathbf{f})(y - \mathbf{v})$$

if not zero $= -(\mathbf{v}-\mathbf{f})^2$ and therefore essentially negative we may write

$$\sum_i (x_i - \mathbf{f})(y_i - \mathbf{v}) < 0$$

to express negative of

$$(x - \mathbf{f})(y - \mathbf{v}) = 0$$

which is the same as

$$\sum_i (x_i - \mathbf{f})(y_i - \mathbf{v}) = 0$$

(Peirce, 1986b, p. 217)

The "negative",

$$\sum_i (x_i - \mathbf{f})(y_i - \mathbf{v}) < 0,$$

means

$$(\exists i)[x(i) \wedge \neg y(i)],$$

which is the negation of

$$(\forall i)[\neg x(i) \vee y(i)],$$

of

$$(\forall i)[x(i) \rightarrow y(i)],$$

and of

"$x$ is a subset of $y$".

Whether or not Peirce knew of Frege's work, his own development of quantification proceeded within a system of ideas completely different from Frege's and on which Peirce had been working for over ten years before the publication of *Begriffsschrift*. Peirce first published his representation theorem in 1875, and the ideas in his 1881 proof of this theorem led directly to the $\sum$ and $\prod$ notation in his 1883 paper.

## 3.5   1885 system of quantificational logic

In this section Peirce's great paper of 1885, 'On the Algebra of Logic: A Contribution to the Philosophy of Notation', will be discussed. There is an inevitably fragmentary or disjointed appearance to the present section because Peirce's own paper is to some extent fragmentary and disjointed. Nevertheless, the paper is filled with important ideas in logic.

### 3.5.1   Quantifiers

In his paper of 1885 Peirce reinterpreted the relation letters with their subscripts as propositions instead of as truth values (1933a, pp. 226–227). In today's terms these subscripted letters can be regarded as atomic formulas. He also reinterpreted $\sum$ and $\prod$ as denoting the *logical* notions of 'for some' and 'for all':

> ... Here, in order to render the notation as iconical as possible we may use $\sum$ for *some*, suggesting a sum, and $\prod$ for *all*, suggesting a product. Thus $\sum_i x_i$ means that $x$ is true of some one of the individuals denoted by $i$ or
>
> $$\sum_i x_i = x_i + x_j + x_k + etc.$$
>
> In the same way, $\prod_i x_i$ means that $x$ is true of all these individuals, or
>
> $$\prod_i x_i = x_i x_j x_k, etc.$$
>
> ... It is to be remarked that $\sum_i x_i$ and $\prod_i x_i$ are only similar to a sum and a product; they are not strictly of that nature, because the individuals of the universe may be innumerable. (Peirce, 1933a, p. 228)

Peirce, in the 1885 paper, overloaded the terms "sum" and "product", in that they may mean the sum and product of truth values, on the one hand, or the disjunction and conjunction of propositions on the other hand.

Hence, we should read the passage above in this way:

> ...Here, in order ...we may use $\sum$ for *some*, suggesting a [disjunction], and $\prod$ for *all*, suggesting a [conjunction]. ...$\sum_i x_i$ and $\prod_i x_i$ are only similar to a [disjunction] and a [conjunction]; they are not strictly of that nature, because the individuals of the universe may be innumerable.

By "innumerable", Peirce probably meant countably infinite here, since that is the way the term is used in his *Brief Description* of 1882 (1933a, p. 180; 1986b, p. 328), although it did mean uncountable in a manuscript of 1893 (1933b, pp. 85–86). Our reading of the passage makes it doubtful that he meant literally infinite disjunctions and conjunctions.

## 3.5.2   Prenex formulas

Peirce did not give a formal definition of his language, especially not by using inductive definitions. In this respect there is a striking difference between Peirce's and Frege's work, which certainly makes it more difficult for the reader to obtain a clear idea of Peirce's system. Nevertheless, he did understand that he was dealing with a language; he said at the start of the 1885 paper:

> The index asserts nothing; it only says "There!" It takes hold of our eyes, as it were, and forcibly directs them to a particular object, and there it stops. Demonstrative and relative pronouns are nearly pure indices, because they denote things without describing them; so are the letters on a geometrical diagram, and the subscript numbers which in algebra distinguish one value from another without saying what those values are. (Peirce, 1933a, p. 211)

> ...The actual world cannot be distinguished from a world of imagination by any description. Hence the need of pronoun and indices, and the more complicated the subject the greater the need of them. (p. 212)

Following the passage introducing the quantifiers, quoted in the previous paragraph above, he gave a number of examples using the language to symbolize first-order statements. As we saw in Section 3.4.1, Peirce used "Quantifying part" to mean the entire string of $\prod$'s and $\sum$'s, with their indices or variables (1933a, p. 227). Here is one of his examples:

> Let $l_{ij}$ mean that $i$ is a lover of $j$, and $b_{ij}$ that $i$ is a benefactor of $j$. Then
>
> $$\prod_i \sum_j l_{ij} b_{ij}$$

means that everything is at once a lover and a benefactor of something; and

$$\prod_i \sum_j l_{ij} b_{ji}$$

that everything is a lover of a benefactor of itself. (Peirce, 1933a, p. 229)

Peirce did have non-prenex formulas in the following passage, but these serve as examples in some rules for combining first-order prenex formulas into a single prenex formula:

> ...the different premisses having been written with distinct indices (the same index not used in two propositions) are written together, and all the $\prod$'s and $\sum$'s are to be brought to the left. This can evidently be done, for

$$\prod_i x_i \cdot \prod_j x_j = \prod_i \prod_j x_i x_j$$

$$\sum_i x_i \cdot \prod_j x_j = \sum_i \prod_j x_i x_j$$

$$\sum_i x_i \cdot \sum_j x_j = \sum_i \sum_j x_i x_j.$$

(Peirce, 1933a, p. 231)

Several more rules for manipulating quantifiers are thereafter given.

### 3.5.3 No material implication sign

Peirce used the Boolean propositional connectives "and", "or", and "not"; he did not make much use of material implication, even though, in his 1870 paper, he used "—<" ambiguously to mean class inclusion, material implication, or logical consequence. In the following quotation from the 1885 paper is the only appearance of the sign between two first-order formulas (Peirce, 1933a, p. 231):

$$\sum_i \prod_j x_i y_j = \prod_j \sum_i x_i y_j$$

...does not hold when the $i$ and $j$ are not separated. We do have, however,

$$\sum_i \prod_j x_i y_j -< \prod_j \sum_i x_i y_j.$$

By this last expression, Peirce meant probably that the right hand formula is a logical consequence of the left hand formula.

### 3.5.4   Second-order logic

Peirce distinguished first- and second-order logic in his 1885 paper, calling them first- and second-intentional logic. Shortly before he introduced the quantifiers, he said:

> The algebra of Boole affords a language by which anything may be expressed which can be said without speaking of more than one individual at a time. It is true that it can assert that certain characters belong to a whole class, but only such characters as belong to each individual separately. The logic of relatives considers statements involving two and more individuals at once. Indices are here required. (Peirce, 1933a, p. 226)

The first-intentional logic provides indices (individual variables) and quantification for individuals, while relation symbols are tokens (constants). In the second-intentional logic, relations may appear as indices and these indices may be quantified or free:

> Our notation, so far as we have developed it, does not show us even how to express that two indices, $i$ and $j$, denote one and the same thing. We may adopt a special token of second intention, say $\mathbf{1}$, to express identity, and may write $\mathbf{1}_{ij}$. But this relation of identity has peculiar properties. The first is that if $i$ and $j$ are identical, whatever is true of $i$ is true of $j$. This may be written
>
> $$\prod_i \prod_j \{\bar{1}_{ij} + \bar{x}_i + x_j\}.$$
>
> The use of the general index of a token, $x$, here, shows that the formula is iconical. (Peirce, 1933a, p. 233)

In other words, $x$ is a free variable for a monadic predicate, and a translation into today's notation would be,

$$(\forall\alpha)(\forall i)(\forall j)[i = j \rightarrow \{\alpha(i) \rightarrow \alpha(j)\}].$$

Note that Peirce's verbal expression in the quotation used material implication, but that his formula used disjunction and negation. He apparently thought in terms of material implications and was quite skilled in translating between such thoughts and formulas in Boolean connectives; but as we mentioned in the previous section above, he did not make much use of a connective for material implication in his language. Although it is simple propositional logic to make such translations today, this would not have been at all easy for his contemporaries. Peirce had become adept at using these logical equivalences from elementary propositional logic, favoring the first two:

$$(\neg p \vee q) \equiv (\neg(p \wedge \neg q)) \equiv (p \rightarrow q).$$

In the next passage, a new token $q$ is introduced for the higher order relation between individuals and predicates:

> ...The other property is that if everything which is true of $i$ is true of $j$, then $i$ and $j$ are identical. This is most naturally written as follows: Let the token, $q$, signify the relation of a quality, character, fact, or predicate to its subject. Then the property we desire to express is
>
> $$\prod_i \prod_j \sum_k (1_{ij} + \bar{q}_{ki}q_{kj}).$$

And identity is defined thus

$$1_{ij} = \prod_k (q_{ki}q_{kj} + \bar{q}_{ki}\bar{q}_{kj}).$$

That is, to say that things are identical is to say that every predicate is true of both or false of both. It may seem circuitous to introduce the idea of a quality to express identity; but that impression will be modified by reflecting that $q_{ki}q_{kj}$ merely means that $i$ and $j$ are both within the class or collection $k$. If we please, we can dispense with the token $q$, by using the index of a token and by referring to this in the Quantifier just as subjacent indices are referred to. That is to say, we may write

$$1_{ij} = \prod_x (x_i x_j + \bar{x}_i \bar{x}_j).$$

(Peirce, 1933a, p. 234)

Or in modern notation,

$$i = j \text{ iff } (\forall \alpha)[\alpha(i) \leftrightarrow \alpha(j)].$$

## 3.6 Logical work after 1885

After 1884, Peirce was cut off from the group of Johns Hopkins students with whom he had worked. The Trustees of Johns Hopkins conspired to eject him from the university, although they arranged to make it appear to be a 'reorganization' of faculty appointments. It is no exaggeration to say that this was disastrous for Peirce's subsequent intellectual life, because the idea of a community of scientific inquirers was central to his philosophy of scientific progress. Christine Ladd-Franklin lamented in a memorial article about Peirce that he had not had students and colleagues to interact with in his later years:

If Charles S. Peirce had happened to have a longer period of activity at the Johns Hopkins University—if the years had not been cut off during which he was kept upon the solid ground of intelligible reason by discussions with a constantly growing group of level-minded students,—there is no doubt that his work would have been of more certain value than it can be affirmed to be now... (Ladd-Franklin, 1916, p. 722)

By 1891, Peirce had retired with his second wife, Juliette, to a house in Milford, Pennsylvania. There they lived in ever increasing poverty and obscurity until Peirce died in 1914. His later papers did not have much influence on the model-theoretic line of development in mathematical logic.

We will now consider two items from his 1896 review of Ernst Schröder's work: the idea of logical consequence and the present-day sense of 'quantifier'. The model-theoretic approach and the idea of logical consequence are implicit in this somewhat strange example[3]:

... suppose the assertion of logical necessity is the assertion that from the proposition, "Some woman is adored by all catholics," it logically follows that "Every catholic adores some woman." That is as much as to say that, for every imaginable set of subjects, either it is false that some woman is adored by all catholics or it is true that every catholic adores some woman. (Peirce, 1933a, p. 278)

By "logical necessity", Peirce meant logical consequence; thus he was investigating the model-theoretic analogue of the deduction theorem. The assertion is,

$$(\exists b)(\forall a)[W(b) \wedge \{C(a) \rightarrow D(a,b)\}] \models$$
$$(\forall a)(\exists b)[W(b) \wedge \{C(a) \rightarrow D(a,b)\}].$$

As Peirce said, "That is as much as to say that, for every imaginable set of subjects,"

$$\models \neg(\exists b)(\forall a)[W(b) \wedge \{C(a) \rightarrow D(a,b)\}] \vee \tag{3.3}$$
$$(\forall a)(\exists b)[W(b) \wedge \{C(a) \rightarrow D(a,b)\}].$$

Continuing with Peirce's argument,

We try the experiment. In order to avoid making it false that some woman is adored by all catholics, we must choose our set of indices so that there shall be one of them, $B$, such that, taking any one, $A$, no matter what, $B$ is a woman, and moreover either $A$ adores $B$ or else $A$ is a non-catholic. (Peirce, 1933a, p. 278)

---

[3]See Section 5.3.1 for a possible explanation of the content of this example.

To restate Peirce's conditions for making the left disjunct in (3.3) false (that is, to *avoid* making the EA statement false), there exists at least one $b$ such that for any $a$,

$$W(b) \text{ and either } D(a,b) \text{ or else } \neg C(a).$$

Here, the Boolean form of the original conjecture makes Peirce's meaning more clear:

$$\models \neg(\exists b)(\forall a)[W(b) \wedge \{\neg C(a) \vee D(a,b)\}] \vee$$
$$(\forall a)(\exists b)[W(b) \wedge \{\neg C(a) \vee D(a,b)\}].$$

To continue,

> But that being the case, no matter what index, $A$, we may take, either $A$ is a non-catholic or else an index can be found, namely, $B$, such that $B$ is a woman, and $A$ adores $B$. (Peirce, 1933a, p. 278)

In other words, if we can falsify the left disjunct in a particular model, then that model will necessarily satisfy the right disjunct; the witness for $b$ in the model will serve as a witness to satisfy the right disjunct for each $a$:

> We see, then, by this experiment, that it is impossible so to take the set of indices that the proposition of consecution shall be false. The experiment may, it is true, have involved some blunder; but it is so easy to repeat it indefinitely, that we readily acquire any desired degree of certitude for the result. (Peirce, 1933a, p. 278)

And so, the idea of logical consequence ("consecution") and the model-theoretic approach are undeniably involved in this passage.

Later in the 1896 review, Peirce used the term "quantifier" in today's sense:

> My general algebra of logic...consists in simply attaching indices to the letters of an expression in the Boolian algebra, making what I term a Boolian, and prefixing to this a series of "quantifiers," which are the letters $\prod$ and $\sum$, each with an index attached to it. (Peirce, 1933a, pp. 282–283)

Hence, the *term* "quantifier", with respect to its present-day sense, was used by Peirce from 1885 on.[4]

It may turn out that Peirce's work on logical graphs will have been an important precursor of some technique in logic or computer science, but such work did not affect Schröder, Löwenheim, or Skolem. These

---

[4]See section 7.3.2, for an earlier usage of "quantified predicate."

graphs are prefigured in Peirce's 1897 review of Schröder's work (1933a, pp. 295–316), and they are given a more thorough exhibition in several articles published in *The Monist* of 1906 and 1908 (1933b, pp. 439–463, 473–537). As Peirce got more and more involved with these graphs, he left alone the first-order logic that we use today. At the 1903 Lowell lectures on pragmatism, he apparently handed out several pages on logic containing a strange attempt to name all of the two quantifier forms according to an extension of medieval or scholastic terminology (1933a, pp. 366–387).

# Chapter 4

# Between Mathematics and Logic

In his Preface to the second (1787) edition of the *Critique of Pure Reason*, Immanuel Kant wrote these well-known words:

> ... since Aristotle [logic] has not [been] required to retrace a single step... It is remarkable also that to the present day this logic has not been able to advance a single step, and is thus to all appearance a closed and completed body of doctrine. (Kant, 1958, p. 17)

But the appearance of being "closed and completed" did not indicate perfection; rather, the traditional logic which immediately preceded the development of mathematical logic was simply trapped in a limited system of ideas that were usually confused with philosophical subjects. Within a century of when Kant wrote the words above, an immensely more powerful logic had been developed independently by Peirce and Frege (taking into account the discussion in Section 3.4.2 above), both of whom were also concerned with its applications to philosophy. And in fact, Leibniz had already introduced a new mathematical approach to logic a century before Kant wrote his *Critique of Pure Reason*, although this approach had only limited success until 1847, when Boole published *A Mathematical Analysis of Logic*.

The term 'mathematical logic' was used at least as early as 1891 by Giuseppe Peano in the title of his paper 'Principii di logica matematica' (1973, pp. 153–161). In present-day usage the term has two distinct senses: it may mean 'logic' treated in a mathematical way, especially using symbols and sometimes called "symbolic logic"; or it may mean the kind (or kinds) of 'logic' used by mathematicians in their proofs. Boole and De Morgan

were concerned with the first sense; Peirce was concerned with both senses—
perhaps even combining them. Peano, in his work, gradually turned the
first sense into the second. (Kennedy, 1973, pp. 7, 75, 101, 153, 190)

The first apprehension of a mathematical kind of logic is usually
attributed to Leibniz:

> As I was keenly occupied with this study, I happened un-
> expectedly upon this remarkable idea, that an alphabet of
> human thought could be devised, and that everything could
> be discovered (*inveniri*) and distinguished (*dijudicari*) by the
> combination of the letters of this alphabet and by the analysis
> of the resulting words.
>
> A *characteristic* of *reason*, by means of which truths would
> become available to reason by some method of calculation, as
> in arithmetic and algebra, so in every other domain, so long as
> it submits to the course of deduction.
>
> Then, in case of a difference of opinion, no discussion between
> two philosophers will be any longer necessary, as (it is not)
> between two calculators. It will rather be enough for them
> to take pen in hand, set themselves to the abacus, and (if it
> so pleases, at the invitation of a friend) say to one another:
> *Calculemus!* (Bochenski, 1970, pp. 274–275)

Leibniz was thus concerned with settling disputes, probably philosophical
and legal ones. In his conception, there is a mathematical logic in the sense
of traditional logic expressed in symbols that are subject to the rules of a
calculus, but there is not yet a mathematical logic in the other sense of a
logic adequate to express mathematical proofs (Scholz, 1961, pp. 50–57).

In this chapter we present Peirce's views on the relation between
mathematics and logic, with some mention of metaphysics and science.
Peirce was very much more concerned with metaphysical questions and
the applicability of logic to such questions than are most mathematical
logicians, which may have contributed to the lack of recognition for his
work. The material in this chapter will serve as evidence that Peirce
had a semantic or model-theoretic attitude towards logic, even though
his attitude did not result in a full formalization of his language and
set-theoretic semantics.

# 4.1   Mathematics, logic, and metaphysics

Whatever it was that Peirce actually meant by "metaphysics" is of merely
incidental importance in this book; but Peirce's lifelong concern with what
he meant is important because that concern seems to have motivated him
to develop a logic that is mathematical in both of the senses just described.

In this chapter and the next, we shall see how deeply Peirce believed that metaphysics is important to the enterprise of science.

Boole introduced a symbolic or mathematical treatment of traditional syllogistic logic in his 1847 monograph, *A Mathematical Analysis of Logic*, as discussed in Section 2.1 above. In it he made this remark about mathematics, logic, and metaphysics:

> I am then compelled to assert, that according to this view of the nature of Philosophy, *Logic forms no part of it*. On the principle of a true classification, we ought no longer to associate Logic and Metaphysics, but Logic and Mathematics. (Boole, 1847, p. 13)

Peirce put himself squarely in Boole's tradition by writing in his 1896 review of Schröder:

> ... we homely thinkers believe that, considering the immense amount of disputation there has always been concerning the doctrines of logic, and especially concerning those which would otherwise be applicable to settle disputes concerning the accuracy of reasonings in metaphysics, the safest way is to appeal for our logical principles to the science of mathematics... (Peirce, 1933a, p. 269)

And in his 1902 manuscript for an unpublished book on logic, Peirce wrote:

> It does not seem to me that mathematics depends in any way upon logic. ... On the contrary, I am persuaded that logic cannot possibly attain the solution of its problems without great use of mathematics. Indeed all formal logic is merely mathematics applied to logic. (Peirce, 1933b, p. 189)

For Peirce, *logic depends on mathematics*, both in the sense that logic must be expressed as a calculus in a mathematical symbolism (in order to be clear and efficacious) and in the sense that the logic or reasoning used by mathematicians should be taken as exemplary (because it has been the most successful). Frege fully characterized in a formal system the logic actually used by mathematicians, and he showed that the traditional logic of syllogism is but a trivial part of his *Begriffsschrift* (1977, pp. 15–17, 51–54); but for him *mathematics depends on logic* in the sense that mathematics is a part of logic and that it should be developed within logic.

As early as 1871, in his review of a new edition of George Bishop Berkeley's writings, Peirce implied that he might use the logic of relations to prove that realism is the true metaphysical basis for science:

> ... it is allowable to suppose that science has no essential affinity with the philosophical views [nominalism] with which it seems

to be every year more associated...Yet a man who enters into the scientific thought of the day and has not materialistic tendencies, is getting to be an impossibility. So long as there is a dispute between nominalism and realism, so long as the position we hold on the question is not determined by any proof *indisputable*, but is more or less a matter of inclination, a man as he gradually comes to feel the profound hostility of the two tendencies will, if he is not less than man, become engaged with one or other... There is, indeed, no reason to suppose the logical question is in its own nature unsusceptible of solution. But that path out of the difficulty lies through the thorniest mazes of a science as dry as mathematics. Now there is a demand for mathematics; it helps to build bridges and drive engines, and therefore it becomes somebody's business to study it severely. But... the faculty of philosophizing, except in the literary way, is not called for; and therefore a difficult question cannot be expected to reach solution until it takes some practical form. If anybody should have the good luck to find out the solution, nobody else would take the trouble to understand it. But though the question of realism and nominalism has its roots in the technicalities of logic, its branches reach about our life. (Peirce, 1958b, pp. 37–38; 1984, pp. 486–487)

For "a science as dry as mathematics" which will solve "the logical question" of realism, Peirce surely had in mind his 1870 logic of relations; his reference to "good luck" implied that he did not yet see how to solve "the logical question". But by "realism" he did not mean that concepts are eternal, intangible entities in some extra-physical realm, which can be examined by the mind's eye—a view that is usually called "Platonism". Peirce spent some pages explaining the realism-nominalism controversy at the beginning of his 1871 review of Berkeley (1958b, pp. 11–25; 1984, pp. 464–476), but he gave the best statement of his definition of "reality" in his 1878 paper, 'How To Make Our Ideas Clear':

> ... an abstract definition of the real ... may perhaps be reached by considering the points of difference between reality and its opposite, fiction. A figment is a product of somebody's imagination; it has such characters as his thought impresses upon it. That whose characters are independent of how you or I think is an external reality. ... Thus we may define the real as that whose characters are independent of what anybody may think them to be. (Peirce, 1934, pp. 265–266; 1986a, p. 271)

> ... all the followers of science are fully persuaded that the processes of investigation, if only pushed far enough, will give one certain solution to every question to which they can be

applied. ... This activity of thought by which we are carried, not where we wish, but to a foreordained goal, is like the operation of destiny. No modification of the point of view taken, no selection of other facts for study, no natural bent of mind even, can enable a man to escape the predestinate opinion. This great law is embodied in the conception of truth and reality. The opinion which is fated to be ultimately agreed to by all who investigate, is what we mean by the truth, and the object represented in this opinion is the real. That is the way I would explain reality. (1934, p. 268; 1986a, p. 273)

In the second of his 1898 Cambridge Conference lectures, Peirce made a somewhat stronger and more precise claim regarding a proof of realism:

Now what was the question of realism and nominalism? I see no objection to defining it as the question of which is the best, the laws or the facts under those laws. ... Roughly speaking, the nominalists conceived the general element of cognition to be merely a convenience for understanding this and that fact and to amount to nothing except for cognition, while the realists, still more roughly speaking, looked upon the general, not only as the end and aim of knowledge, but also as the most important element of being. (Peirce, 1933b, p. 3)

... My plan for defeating nominalism is not simple nor direct; but it seems to me sure to be decisive, and to afford no difficulties except the mathematical toil that it requires. For as soon as you have once mounted the vantage-ground of the logic of relatives ... you find that you command the whole citadel of nominalism, which must thereupon fall almost without another blow. (p. 5)

Thus Peirce seems to have planned for many years to pursue metaphysics through the method of science by the use of a logic developed mathematically from mathematical reasoning. We take up this topic again in Section 5.3.4, where we consider Peirce's promise to give a "proof" of his concept of pragmatism.

## 4.2  Peirce on mathematics and logic

A preview of Peirce's views was given in Chapter 1, Section 1.4, above. The following account is based on two of Peirce's writings: In 'The Regenerated Logic', Peirce's 1896 review of Schröder's *Exact Logic*, he described how mathematics and logic are related to each other and to the other sciences (1933a, pp. 266–272). In 'The Simplest Mathematics', Peirce's

1902 manuscript for Chapter 3 of his unpublished book, *Minute Logic*, he gave definitions of mathematics and logic, further explaining the difference and proper relation between them (1933b, pp. 189–206). Regardless of the extent to which one can agree with Peirce, his views demonstrate a deep understanding of mathematics and logic as they are practiced at the level of original research.

## 4.2.1 The success of mathematics; logic and the sciences

In Peirce's review of Schröder, already discussed in Sections 1.4 and 3.6, he said that mathematics is the most abstract science because it deals exclusively in hypotheses. The mathematician observes examples or schemata in thought and makes inductions from them. Even though a mathematician's thought is fallible, it is easy to repeat these thought experiments and to make new ones:

> ... questions of philosophy ought to be decided upon logical principles, these having been themselves settled upon principles derived from the only science in which there has never been a prolonged dispute relating to the proper objects of that science. (Peirce, 1933a, p. 266)

> It is a remarkable historical fact that there is a branch of science in which there has never been a prolonged dispute concerning the proper objects of that science. It is the [*sic*] mathematics. Mistakes in mathematics occur not infrequently, and not being detected give rise to false doctrine, which may continue a long time. ... Again, there have been prolonged disputes among real mathematicians concerning questions which were not mathematical or which had not been put into mathematical form. ... But the fact remains that concerning strictly mathematical questions, and among mathematicians who could be considered at all competent, there has never been a single prolonged dispute.

> It does not seem worth while to run through the history of science for the sake of the easy demonstration that there is no other extensive branch of knowledge of which the same can be said.

> ... the reason for this immunity of mathematics ... arises from the fact that the objects which the mathematician observes and to which his conclusions relate are objects of his mind's own creation. Hence, although his proceeding is not infallible—which is shown by the comparative frequency with which mistakes

are committed and allowed—yet it is so easy to repeat the inductions upon new instances, which can be created at pleasure, and extreme cases can so readily be found by which to test the accuracy of the processes, that when attention has once been directed to a process of reasoning suspected of being faulty, it is soon put beyond all dispute either as correct or as incorrect. (pp. 268–269)

The success of mathematics in rapidly settling disputes is in part because it lacks any need for expensive equipment, such as chemical apparatus or superconducting super-colliders, or any need for distant journeys to remote locations on earth or to outer space. Mathematicians have developed logic to the highest state insofar as there are the fewest obstacles to its rapid and effective exercise. The success of mathematics is also in part attributable to the reality of its subject matter: For Peirce, mathematics was real in the sense given in the previous section above; it does not depend on what anyone wishes or believes, and its effects are all around us in the form of scientific and technological achievements. Mathematicians do not 'make up' mathematics in the way that poets and painters make up poems and paintings, nor in the way that philosophers make up systems of metaphysics; mathematicians 'make up' mathematics in the way that physicists and chemists make up experiments in the laboratory. Hence, disputes are rapidly settled because competent mathematicians who do the same 'experiment' obtain the same results. Logic must become 'real' in the way that mathematics is real, but its effects are not so easily conceived or apprehended as are the concrete effects of mathematics.

> Hence, ... considering the immense amount of disputation there has always been concerning the doctrines of logic, and especially concerning those which would otherwise be applicable to settle disputes concerning the accuracy of reasonings in metaphysics, the safest way is to appeal for our logical principles to the science of mathematics, where error can only long go unexploded on condition of its not being suspected. (Peirce, 1933a, p. 269)

Thus, for Peirce logic is to be based on mathematics in the sense that the principles of logic are to be fixed by examining mathematical thought and mathematical reasoning. This logic can then be applied to philosophy and the other less abstract sciences. Here we must emphasize that Peirce's view is the opposite of Frege's logicist view that mathematics is to be based on logic. To continue,

> This double assertion, first, that logic ought to draw upon mathematics for control of disputed principles, and second, that ontological philosophy ought in like manner to draw upon logic, is a case under a general assertion ... that the sciences

71

may be arranged in a series with reference to the abstractness of their objects; and that each science draws regulating principles from those superior to it in abstractness, while drawing data for its inductions from the sciences inferior to it in abstractness... (Peirce, 1933a, pp. 269–270)

Peirce then illustrated this idea with a table of dependency relating mathematics, philosophy (namely, logic and metaphysics), and the sciences:

| MATHEMATICS | | |
|---|---|---|
| Philosophy $\left\{\begin{array}{l}\text{Logic}\\\text{Metaphysics}\end{array}\right.$ | | |
| Science of Time | Geometry | |
| Nomological Psychics | Nomological Physics | $\left\{\begin{array}{l}\text{Molar}\\\text{Molecular}\\\text{Ethereal}\end{array}\right.$ |
| Classificatory Psychics | Classificatory Physics | $\left\{\begin{array}{l}\text{Chemistry}\\\text{Biology, or the}\\\text{chemistry of}\\\text{protoplasms}\end{array}\right.$ |
| Descriptive Psychics | Descriptive Physics | |
| PRACTICAL SCIENCE | | |

Table 4.1: Mathematics, philosophy, and the sciences
(Peirce, 1933a, p. 270)

The sciences above depend upon the sciences below. Note that all but mathematics and logic depend on metaphysics. Thus, for Peirce metaphysics was important to science, but it had not yet been developed scientifically. After giving the table, Peirce continued,

... only the first three branches concern us here.

Mathematics is the most abstract of all the sciences. For it makes no external observations, nor asserts anything as a real fact. When the mathematician deals with facts, they become for him mere "hypotheses"; for with their truth he refuses to concern himself. The whole science of mathematics is a science of hypotheses; so that nothing could be more completely abstracted from concrete reality. Philosophy is not quite so abstract. ... philosophy, in the strictest sense, confines itself to such observations as *must* be open to every intelligence which can learn from experience. ...logic is much more abstract even than metaphysics. For it does not concern itself with any facts not implied in the supposition of an unlimited applicability of language.

Mathematics is not a positive science; for the mathematician holds himself free to say that *A* is *B* or that *A* is not *B*, the only obligation upon him being, that as long as he says *A* is *B*, he is to hold to it, consistently. But logic begins to be a positive science; since there are some things in regard to which the logician is not free to suppose that they are or are not; but acknowledges a compulsion upon him to assert the one and deny the other. Thus, the logician is forced by positive observation to admit that there is such a thing as doubt, that some propositions are false, etc. But with this compulsion comes a corresponding responsibility upon him not to admit anything which he is not forced to admit.

Logic may be defined as the science of the laws of the stable establishment of beliefs. Then, exact logic will be that doctrine of the conditions of establishment of stable belief which rests upon perfectly undoubted observations and upon mathematical, that is, upon *diagrammatical*, or, *iconic*, thought. We, who are sectaries of "exact" logic, and of "exact" philosophy, in general, maintain that those who follow such methods will, so far as they follow them, escape all error except such as will be speedily corrected after it is once suspected. (Peirce, 1933a, pp. 270–271)

"Diagrammatical" refers to the algebraic or geometric diagrams used by mathematicians in the activity of proving theorems; "iconic" means that individuals, relations, and operations are represented by constants, variables, and other symbols. Thus, the second sense of 'mathematical logic' has emerged: besides being a *mathematical treatment* of traditional logic, using symbols especially from algebra, a mathematical logic must also be based on *the kind of reasoning* used by mathematicians, from which Peirce proposed to develop a logic applicable to all the sciences. Furthermore, the two senses are united insofar as this kind of logic or reasoning used by mathematicians is to be expressed in symbols that are subject to the rules of a calculus; the symbols are not to be merely the static abbreviations of words as was the case in ancient and medieval symbolizations, and in De Morgan's notation for relations.

Toward the end of his 1896 review of Schröder, Peirce predicted some future successes for his development of logic:

Of what use does this new logical doctrine promise to be? The first service it may be expected to render is that of correcting a considerable number of hasty assumptions about logic which have been allowed to affect philosophy. In the next place, if Kant has shown that metaphysical conceptions spring from formal logic, this great generalisation upon formal logic must lead to a new apprehension of the metaphysical conceptions which shall render them more adequate to the needs of science. In short, "exact" logic will prove a stepping-stone to "exact" metaphysics. In the next place, it must immensely widen our logical notions. ... In the next place, ... objective logic, mentioned at the beginning of this article, is destined to grow into a colossal doctrine which may be expected to lead to most important philosophical conclusions. Finally, the calculus of the new logic, which is applicable to everything, will certainly be applied to settle certain logical questions of extreme difficulty relating to the foundations of mathematics. Whether or not it can lead to any method of discovering methods in mathematics it is difficult to say. Such a thing is conceivable. (Peirce, 1933a, p. 286)

Again, note the importance to Peirce of metaphysics as a science which needs the application of a mathematical logic in order to rise above its moribund state of endless disputes. If metaphysics were to become scientific, it would progress positively through the identification and solution of problems in the same way that mathematics progresses through the conjecture and proof of theorems. By the term "foundations of mathematics", he probably meant set theory, which he had begun to study around 1889. Peano had used the term in 1889 with reference to the principles of arithmetic (1973, p. 101). And the word "conceivable" at the end of the quotation above is also significant, as we shall see in the next chapter, because the notion of "conceivable effects" was central to Peirce's concept of pragmatism.

## 4.2.2  Defining mathematics and logic

At the start of the 1902 manuscript for his unpublished book on *Minute Logic*, Peirce gave his father's definition of the method of mathematics and his own definition of its subject matter:

It was Benjamin Peirce, whose son I boast myself, that in 1870 first defined mathematics as "the science which draws necessary conclusions." This was a hard saying at the time; but today, students of the philosophy of mathematics generally acknowledge its substantial correctness. (Peirce, 1933b, p. 189)

...For all modern mathematicians agree with Plato and Aristotle that mathematics deals exclusively with hypothetical states of things, and asserts no matter of fact whatever; and further, that it is thus alone that the necessity of its conclusions is to be explained. This is the true essence of mathematics; and my father's definition is in so far correct that it is impossible to reason necessarily concerning anything else than a pure hypothesis. (pp. 191–192)

Mathematics is the study of what is true of hypothetical states of things. That is its essence and definition. (p. 193)

It is difficult to decide between the two definitions of mathematics; the one by its method, that of drawing necessary conclusions; the other by its aim and subject matter, as the study of hypothetical states of things. The former makes or seems to make the deduction of the consequences of hypotheses the sole business of the mathematician as such. But it cannot be denied that immense genius has been exercised in the mere framing of ... general hypotheses ... Even the framing of the particular hypotheses of special problems almost always calls for good judgment and knowledge, and sometimes for great intellectual power... (pp. 197–198)

Peirce then gave a definition for the method of logic and contrasted it with his father's definition for the method of mathematics:

The philosophical mathematician, Dr. Richard Dedekind, holds mathematics to be a branch of logic. This would not result from my father's definition, which runs, not that mathematics is the science of drawing necessary conclusions—which would be deductive logic—but that it is the science which draws necessary conclusions. It is evident, and I know as a fact that he had this distinction in view. (Peirce, 1933b, pp. 198–199)[1]

The complementarity of these definitions was then exemplified in a description of the concerns of the mathematician versus the concerns of the logician, based on Peirce's own work with his father on linear associative algebras and the logic of relations, work which resulted in their two papers of 1870:

At the time when he thought out this definition, he, a mathematician, and I, a logician, held daily discussions about a large subject which interested us both; and he was struck, as I was, with the contrary nature of his interest and mine in the same

---

[1]Peirce gave a footnote to the forward of *Was sind und was solen die Zahlen* (Dedekind, 1963, pp. 31–40).

propositions. The logician does not care particularly about
this or that hypothesis or its consequences, except so far as
these things may throw a light upon the nature of reasoning.
The mathematician is intensely interested in efficient methods of reasoning, with a view to their possible extension to
new problems; but he does not, *quâ* mathematician, trouble
himself minutely to dissect those parts of this method whose
correctness is a matter of course. The different aspects which
the algebra of logic will assume for the two men is instructive
in this respect. The mathematician asks what value this algebra has as a calculus. Can it be applied to unravelling a
complicated question? Will it, at one stroke, produce a remote
consequence? The logician does not wish the algebra to have
that character. On the contrary, the greater number of distinct
logical steps, into which the algebra breaks up an inference,
will for him constitute a superiority of it over another which
moves more swiftly to its conclusions. He demands that the
algebra shall analyze a reasoning into its last elementary steps.
Thus, that which is a merit in a logical algebra for one of these
students is a demerit in the eyes of the other. The one studies
the science of drawing conclusions, the other the science which
draws necessary conclusions. (Peirce, 1933b, p. 199)

Here we might recall the fable of the centipede who, when asked how he
could walk using so many legs, got hopelessly confused and tangled up as
he tried to think about how he actually did it.
   Peirce then gave a definition for the subject matter of logic:

But, indeed, the difference between the two sciences is far more
than that between two points of view. Mathematics is purely
hypothetical: it produces nothing but conditional propositions.
Logic, on the contrary, is categorical in its assertions. True it
is not merely, or even mainly, a mere discovery of what really
is, like metaphysics. It is a normative science. ... There is a
mathematical logic, just as there is a mathematical optics and
a mathematical economics. Mathematical logic is formal logic.
Formal logic, however developed, is mathematics. Formal logic,
however, is by no means the whole of logic ... Logic depends
upon mathematics; still more intimately upon ethics; but its
proper concern is with truths beyond the purview of either.
(Peirce, 1933b, pp. 199–200)

For Peirce, "Logic depends upon mathematics" in both senses of the term
'mathematical logic' that we have been discussing: logic needs to have
the symbolical, diagrammatical techniques of mathematics applied to it
in order for it to advance, but it also needs to apply those techniques to

the example of mathematical reasoning in order to have the highest goal *towards* which to advance.

The question then arises, "For whom will this new logic be useful?" Mathematicians seem to have gotten along quite well without it:

> ...mathematics ...has no need of any appeal to logic. No doubt, some reader may exclaim in dissent to this, on first hearing it said. Mathematics, they may say, is preëminently a science of reasoning. So it is; preëminently a science that reasons. But just as it is not necessary, in order to talk, to understand the theory of the formation of vowel sounds, so it is not necessary, in order to reason, to be in possession of the theory of reasoning. Otherwise, plainly, the science of logic could never be developed. (Peirce, 1933b, p. 200)

Hence, those who are most skilled in necessary reasoning do not need to have a theoretical account of their skill in order to practice it successfully— the centipede does not need to be taught a theory of walking in order to walk. This is probably how Peirce would account for the fact that many present-day mathematicians do not take much interest in logic. Mathematics was successful for centuries prior to the development of a science of mathematical logic. And logic did not have a deep enough subject matter until mathematical reasoning and symbolic techniques had been highly developed as practical arts.

> ...if the study of electricity had been pursued resolutely, even if no special attention had ever been paid to mathematics, the requisite mathematical ideas would surely have been evolved. Faraday, indeed, did evolve them without any acquaintance with mathematics. Still it would be far more economical to postpone electrical researches, to study mathematics by itself, and then to apply it to electricity, which was Maxwell's way. In this same manner, the various logical difficulties which arise in the course of every science except mathematics, ethics, and logic, will no doubt get worked out after a time, even though no special study of logic be made. But it would be far more economical to make first a systematic study of logic. (Peirce, 1933b, p. 201)

And so, the new logic will conceivably be useful to metaphysics and to all the sciences, which is why it appears above everything except mathematics in the chart discussed in the previous section.

Finally, without being anachronistic, we can say that Peirce was looking at mathematics the way model theory looks at mathematics today:

> Each branch of mathematics sets out from a general hypothesis of its own. I mean by its general hypothesis the substance of

its postulates and axioms, and even of its definitions, should they be contaminated with any substance, instead of being the pure verbiage they ought to be. We have to make choice, then, between a division of mathematics according to the matter of its hypothesis, or according to the forms of the schemata of which it avails itself. These latter are either geometrical or algebraical. (Peirce, 1933b, p. 204)

... The primary division of mathematics into algebra and geometry is the usual one. It remains ... now that everybody knows that any mathematical subject ... may be treated either algebraically or geometrically...

Let us, then divide mathematics according to the nature of its general hypotheses, taking for the ground of primary division the multitude of units, or elements, that are supposed; and for the ground of subdivision that mode of relationship between the elements upon which the hypotheses focus the attention. (p. 205)

By hindsight, one can see the idea of a first-order theory here. Admittedly, Peirce did not give a rigorous specification either of a formal language or of its interpretation via set-theoretic semantics. Nevertheless, his informality notwithstanding, each branch has its own proper axioms and definitions, informal set-theoretic semantics does provide the interpretation of individual elements and relations, and first-order (or "first-intentional", as Peirce called it) logic expresses the facts about the individual models of a particular theory or hypothesis.

In the logicist program, as exemplified in the writings of Frege, mathematical objects are to be manufactured out of concepts, and logic supplies the means for developing the theory of concepts. Moreover, this logic is developed within an artificial language or *Begriffsschrift* (Frege, 1977). We can compare the logicist program of Frege with the pragmaticist program of Peirce thusly:

**Frege's logicist program:**
- mathematical objects are manufactured out of concepts,
- logic is the means to develop the theory of concepts,
- logic itself is developed in an artificial language.

**Peirce's pragmaticist program:**
- logical principles are derived from mathematical thought,
- logic is the means to develop philosophy and the sciences,
- logical work is carried out in a symbolic calculus.

For various reasons, the views of the logicist school have dominated accounts of the history of mathematical logic: If logicism is not promoted and extolled outright, then, at the very least, the logicist interpretation of the crucial terms is assumed for the sake of historical inquiry. For example, Peirce's logic has been investigated by, first, attempting to translate his notations into the notation of *Principia Mathematica*, and then raising questions as to how many (or how few) anticipations of *Principia Mathematica* can be found in his writings. Given such a context, it is understandable that the work of Boole, De Morgan, Peirce, and Schröder is often dismissed as an early "algebra of logic"—a dead end. In fact, Peirce's philosophy of mathematics and logic was so totally antithetical to logicism that today it is difficult, if not impossible, to understand his work until we interpret crucial terms, such as 'logic' and 'mathematics', in a sympathetic way. But what is and was totally inexcusable and utterly unforgivable is to claim that the so-called "algebra of logic" was identical with the monadic predicate calculus (viz., the syllogistic logic of Aristotle)! Peirce was the first to introduce an iconic notation for binary relations, and to deny this is supreme, willful ignorance. The dialogue in Chapter 6 will provide more context for this point.

Frege's work had one great advantage over Peirce's. Frege presented a formal system in clear definitions; Peirce's system is presented in only a fragmentary way via examples. Therefore, it takes much more work to establish what Peirce's syntax was, and a number of aspects remain unclear even after a great deal of work. Nevertheless, allowing for trivial typographical changes and making use of a material implication sign, Peirce's notation is far closer than Frege's to the notation used today.

Perhaps Peirce hoped that the development of logic would be useful to at least some mathematicians when he predicted at the end of the 1896 review, "Finally, the calculus of the new logic, which is applicable to everything, will certainly be applied to settle certain logical questions of extreme difficulty relating to the foundations of mathematics." (1933a, p. 286) Whatever Peirce actually conceived of when he made this prediction, the Löwenheim–Skolem Theorem, which was developed in a direct line of influence passing from Peirce through Schröder to Löwenheim and then to Skolem, was undeniably the solution to a logical question "of extreme difficulty relating to the foundations of mathematics." This line of influence will be discussed in Chapter 6.

# Chapter 5

# Between Theory and Practice

In 1878, Peirce published the following pragmatic maxim, to which he adhered for the rest of his life:

> Consider what effects, which might conceivably have practical bearings, we conceive the object of our conception to have. Then, our conception of these effects is the whole of our conception of the object. (Peirce, 1934, p. 258; 1986a, p. 266)

We are concerned in the present chapter both to explain what Peirce meant by "pragmatism" and to give some history of the usage of this word: In Section 5.1 we discuss Peirce's maxim or concept of pragmatism; in Section 5.2 we explain the ancient and present-day popular, non-technical usages of the word; and in Section 5.3 we give a historical account of the development of Peirce's concept, including his reaction to the appropriation of the word by others.

In Chapter 1, Section 1.3, we considered the dangers inherent in cutting the Peircean Knot into pieces and reassembling those pieces into a possibly distorted version of Peirce's actual thought. As we remarked in Section 1.1, this book is written in the spirit of van der Waerden: Arguments are given and supported by passages from the original sources. But the organization of the following account required that a number of passages—especially from Peirce's paper, 'What Pragmatism Is'—be considered out of their original sequence. Therefore, we ask the reader to treat this account as a commentary on Peirce's thought and not in any way as a replacement for his original writings. The present chapter could perhaps have been titled, "Peirce's philosophy of science", or "Peirce's philosophy of pragmatism"; instead the title is, "What is the relation between theory and practice?" because Peirce's views were based on his extensive experience as a research scientist, and our purpose is to explain his views rather than to discuss issues in present-day academic philosophy.

## 5.1   Peirce's concept of pragmatism

Peirce wrote in his definition of "Pragmatic and Pragmatism", for J. M. Baldwin's 1902 *Dictionary of Philosophy and Psychology*:

> The opinion that metaphysics is to be largely cleared up by the application of the following maxim for attaining clearness of apprehension: "Consider what effects, that might conceivably have practical bearings, we conceive the object of our conception to have. Then, our conception of these effects is the whole of our conception of that object."
>
> ...This maxim was first proposed by C. S. Peirce in the *Popular Science Monthly* for January, 1878 (xii. 287); and he explained how it was to be applied to the doctrine of reality. The writer was led to the maxim by reflection upon Kant's *Critic of the Pure Reason*. (Peirce, 1934, p. 1)

Peirce began his 1905 article, "What Pragmatism Is", by explaining the source of the pragmatic habit of mind:

> ...every master in any department of experimental science, has had his mind moulded by his life in the laboratory to a degree that is little suspected. ...excepting perhaps upon topics where his mind is trammelled by personal feeling or by his bringing up, his disposition is to think of everything just as everything is thought of in the laboratory, that is, as a question of experimentation. ...whatever assertion you may make to him, he will either understand as meaning that if a given prescription for an experiment ever can be and ever is carried out in act, an experience of a given description will result, or else he will see no sense at all in what you say.
>
> That laboratory life did not prevent the writer ...from becoming interested in methods of thinking; and when he came to read metaphysics, although much of it seemed to him loosely reasoned and determined by accidental preposessions, yet in the writings of some philosophers, especially Kant, Berkeley, and Spinoza, he sometimes came upon strains of thought that recalled the ways of thinking of the laboratory, so that he felt he might trust to them; all of which has been true of other laboratory-men.
>
> Endeavoring, as a man of that type naturally would, to formulate what he so approved, he framed the theory that a

*conception*, that is, the rational purport of a word or other expression, lies exclusively in its conceivable bearing upon the conduct of life; so that, since obviously nothing that might not result from experiment can have any direct bearing upon conduct, if one can define accurately all the conceivable experimental phenomena which the affirmation or denial of a concept could imply, one will have therein a complete definition of the concept and *there is absolutely nothing more in it.* For this doctrine he invented the name *pragmatism.* (Peirce, 1934, pp. 272–274)

In Chapter 4 we saw that for Peirce mathematics, too, is an experimental science—notwithstanding that the subject matter is intangible—because mathematicians can verify each other's experiments. Since Peirce's concept of experiment was not limited to physical experiments, he was not so extreme as the logical positivists were later to be: He believed that a positive science of metaphysics could be developed by means of the method of science applied through a logic derived from mathematical reasoning and expressed in mathematical symbolism as a calculus.

## 5.2 Older usages of "pragmatism"

Although the word "pragmatism" is well-known today, Peirce's concept is hardly known at all. As Philip Wiener writes,

We cannot simply equate the "pragmatic" with the "practical" as is so commonly done by popular writers. ... "practical" in ordinary discourse is often synonymous with the "convenient," the "useful," and the "profitable" and thus contributes to enormous misunderstandings of the serious aims of pragmatism. (P. Wiener, 1973, p. 553)

The idea of skill in practical matters was a central one to the philosophers of ancient Greece, especially to Plato and Aristotle. H. S. Thayer has this to say in *Meaning and Action: A Critical History of Pragmatism,* his major book on the subject:

'Pragmatism,' said William James, was a new name for some old ways of thinking. But there is reason for doubting the propriety of James's comment. In some of its senses—senses to be distinguished shortly—*pragmatism* might with equal justice have been said to be an old name for a new way of thinking.

Etymology supports this cavil. *Pragma* is the Greek for *things, facts, deeds, affairs,* or as James himself notes, "meaning action, from which our words 'practice' and 'practical' come." The

> word is often so used by Plato and Aristotle; but it has had
> another sort of career and made other appearances as well.
> ...In Germany, one expression for a practical-minded person is
> *ein pragmatischer Kopf.* And in Germany, too, in the nineteenth
> century, there was developed what was called the "pragmatic
> method in history" concerned with the practical consequences
> and uses of the study of history. But the notion goes far back
> to the Greek historian Polybius (204–122 B.C.), who liked to
> regard his *Histories* as instructive and useful to the living and,
> accordingly, often referred to his work as *pragmatike historia.*
> (Thayer, 1981a, pp. 7–8)

And according to Wiener,

> The very term "pragmatic" with its Greek root *pragma* ("af-
> fair, practical matter") was borrowed by the Romans to mean
> "skilled in business, and especially, experienced in matters of
> law"; hence a *pragmaticus* was "one skilled in the law, who fur-
> nished orators and advocates with the principles on which they
> based their speeches" (Cicero, *Orationes* 1, 59...). (P. Wiener,
> 1973, p. 554)

Even this ancient sense concerned the relation of theory to practice, insofar as skill in practical matters is the sign of a kind of intuitive theory.
Peirce made that relation explicit and precise in his concept. Thus the
word "pragmatic" was adopted by Peirce for his own special meaning,
but it was certainly not invented by him. Peirce's choice of the word was
influenced mainly by Kant's technical use of the German words *praktisch*
and *pragmatisch,* as we shall see in the next section.

Nowadays, journalists often apply the word "pragmatic" to expedient,
non-ideological, and even unprincipled individuals or actions—especially
to politicians and politics. This present-day usage of the word to describe
politicians and their actions opposes it, implicitly or explicitly, to "ideological": Pragmatic political leaders are flexible, willing to negotiate, and
usually seem to be motivated by a desire to survive as leaders; ideological
political leaders are inflexible, adamant, and usually seem to be motivated
by the dogmatic assurance that they have God or history on their side.
In such a usage, the profound relation between theory and practice that
Peirce sought to explain and to champion is totally lost. To be pragmatic
in Peirce's sense is to be scientific: One acts according to theoretical
principles but also maintains an attitude of fallibilism, so that theory is
continually improved by practice just as practice is reciprocally directed
by theory; one tries always to keep in mind all the conceivable effects
of actions taken, and one certainly does not limit one's thought to the
particular effect merely of staying in power. Thus the word "pragmatism"

is now debased as compared with Peirce's concept, and it has even lost most of the positive content in its ancient meaning.

It is the confusion of conceivable (that is, general) effects with an ever narrowing scope of actual (that is, particular) effects that debases the word with respect to the meaning given to it in Peirce's maxim. The worst of such debasement lies in the casual use of the word to mean crass opportunism.

## 5.3   Development of Peirce's pragmatism

In this section we give a historical account in five stages: first, the Cambridge 'Metaphysical Club' and Peirce's role in its inquiries; second, Peirce's choice of "pragmatism" as the name for his technical concept of the relation between theory and practice in the method of science and for the concept's wider application to the range of subjects discussed in the Club; third, the appropriation and popularization of the word by James and the differences between Peirce's and James's views; fourth, Peirce's reaction to the ensuing confusion as the word seemed to lose all meaning—such a confusion that Peirce coined a new word, "pragmaticism", for his own concept; and finally, we bring the account to a close in the period shortly after Peirce died and make some concluding remarks. For the convenience of the reader, a chronology of Peirce and his concept of pragmatism is given in Appendix B.

### 5.3.1   The Metaphysical Club

Max Fisch describes the beginnings of the Metaphysical Club in his article about Oliver Wendell Holmes and the Club:

> In a long letter from Berlin in 1868 [William] James had written to Holmes: "When I get home let's establish a philosophical society to have regular meetings and discuss none but the very tallest and broadest questions." The society was organized in 1869. The most significant fact about it is that of its six most active members three were lawyers—Holmes, Nicholas St. John Green, and Joseph Warner. The other three were experimental scientists—James, Peirce, and Chauncey Wright. To the three lawyers I venture to add a fourth, John Chipman Gray, not named by Peirce, but mentioned in James's letter proposing the organization. He and Holmes often called together on James during its most active years (1869–1872), and James lists him with Peirce and Holmes among those with whom he gossiped most on generalities. Wright and Green were the natural leaders of the group by right of philosophical maturity. They were both about forty; Peirce, Gray, Holmes, and James

about thirty; Warner in his twenties. The one thing that all seven had in common, besides a Harvard degree, was an enthusiasm for the British tradition in philosophy, and a sense of the epoch-making importance of Darwin's *Origin of Species,* which had appeared a decade before their first meetings. Peirce alone of their number "had come upon the threshing floor of philosophy through the doorway of Kant," and even his ideas were acquiring the British accent.

According to the usual account, the name "pragmatism" was suggested by Kant, the doctrine by reflection on the methods of the experimental sciences in the light of British empiricism. Whatever assertion you make to an experimentalist, said Peirce, "he will either understand as meaning that if a given prescription for an experiment ever can be and ever is carried out in act, an experience of a given description will result, or else he will see no sense at all in what you say" (*CP* 5.411). I suggest, however, that the methods of the practising lawyer had quite as much to do with it. Peirce in fact professed to have done no more than follow the lead of one of the lawyers in the group, "a marvelously strong intelligence," Nicholas Green. (Fisch, 1986b, pp. 8–9)[1]

Peirce's contribution to an 1877 obituary for Green included these words:

> . . . The basis of his philosophy was, that every form of words that means any thing indicates some sensible fact on the existence of which its truth depends. You can hardly call this a doctrine: it is rather an intellectual tendency. But it was Green's mission to insist upon it and to illustrate it. . . . one did not at first so much note his delicate appreciation of what was real, as his scorn for all that was unreal. (Peirce, 1986a, pp. 209–210)

And in a 1906 description of the Club, Peirce wrote,

> . . . Green was one of the most interested fellows, a skillful lawyer and a learned one, a disciple of Jeremy Bentham. His extraordinary power of disrobing warm and breathing truth of the draperies of long worn formulas, was what attracted attention to him everywhere. In particular, he often urged the importance of applying Bain's definition of belief, as "that upon which a man is prepared to act." From this definition, pragmatism is scarce more than a corollary; so that I am

---

[1] Fisch's reference to " *CP* 5.411" denotes (Peirce, 1934, p. 273), which was quoted in Section 5.1 above.

disposed to think of him as the grandfather of pragmatism.
(Peirce, 1934, pp. 7–8)

Fisch, in his article on Alexander Bain and the Metaphysical Club, establishes that Peirce meant: that Green was the grandfather, and Peirce the father, of pragmatism (Fisch, 1986b, p. 82); and that Green had brought the importance of Bain's definition to the attention of the Club, all of whose members were already familiar with Bain's writings (Fisch, 1986b, p. 93). Peirce mentioned Bain in a lecture manuscript as early as 1866 (1958a, p. 352; 1982, p. 495), and in 1870 he reviewed Bain's book on logic (1984, pp. 441–444). Green influenced Wright and Holmes to apply Bain's definition of belief to a theory of cognition and to a prediction theory of law, respectively (Fisch, 1986b, pp. 98–99).

Thayer, in his 1981 article, 'Pragmatism: A Reinterpretation of the Origins and Consequences', points out that the meetings of the Metaphysical Club took place in a social milieu in which progress itself seemed to be threatened. Three social crises in the United States during this period were especially related to failures of the law to prevent or to solve certain social problems: slavery and the Civil War, central banking and inflation, and scandalous corruption of public officials (Thayer, 1981b, pp. 9–13). The first two crises involved conflicting interpretations of the meaning of the U.S. Constitution, and they seemed to indicate that the law fails to provide a method for resolving such conflicts within its own realm. The third crisis involved the apparent failure of the law to bring about certain desired results or to prevent the grossest of crimes. Thayer writes,

> So construed as a method of critical clarification exemplified in the sciences, pragmatism is a reflection upon already-existing procedures, not a recommendation for scientific practice. But the question is: For whom is the recommended remedial method of clarification intended: More precisely, what intellectual forms and activities in American experience in the late nineteenth century appeared to generate the need for pragmatic clarification of ideas? Granted, clarity is a good thing; but why this special philosophical concentration on it in the 1870s? (Thayer, 1981b, pp. 5–6)

> ...My point is only that the special interest in an effective method for clarifying beliefs and the analysis of concepts in the club, circa 1872, was *partially* stimulated by powerful divisive and conflicting interests and activities occurring and evident on the historical scene in the late nineteenth century. And this is to suggest at most that the impulse then to develop a method of clarifying ideas, claims, and policies was generated from more than purely academic philosophical considerations... (p. 16)

Of course, as we have seen in Chapter 4 and shall see in the rest of the present chapter, Peirce himself was very concerned with clarifying ideas in metaphysics.

To Thayer's list of three crises affecting the discussions of the Metaphysical Club should be added at least two more influences: First, Darwin's theory of evolution had a tremendous effect on the scientific community and on society at large, as Fisch mentioned in the passage quoted at the beginning of this section. See *Evolution and the Founders of Pragmatism* (P. Wiener, 1949) [2]. And second, Pope Pius IX's assertion of authority in Roman Catholic doctrinal matters, including his condemnation of Darwin's theory and most of science in general, had a major effect on Christianity and on society at large. To quote from *The New Columbia Encyclopedia*:

> **Pius IX**, 1792–1878, pope (1846–78)...In 1854, Pius declared the dogma of the Immaculate Conception of the Virgin to be an article of faith. In 1864 he issued the encyclical *Quanta cura*, accompanied by a list (*Syllabus*) of erroneous modernistic statements. In 1869 he convoked the First VATICAN COUNCIL, the principal work of which was the enunciation of papal infallibility. Pius IX's pontificate—the longest in history—helped define the role of Roman Catholicism in the modern world. (Harris and Levey, 1975, p. 2159).

These three doctrinal documents of Pius IX shocked and outraged the Protestant and Orthodox Christian world. The theme of church versus state in Dostoevsky's *Brothers Karamozov*, especially in the chapter on 'The Grand Inquisitor', was one response (1990). The Ku Klux Klan was originally organized in 1866 to protect white supremacy, and it was thus a part of the milieu of social crisis that Thayer described; nevertheless, it was reorganized in 1915 not only as a racist organization but also as an anti-Catholic, anti-immigrant one (Harris and Levey, 1975, p. 1505). Later in this section, we quote Peirce's reference to Pius IX with respect to the "most perfect example in history" of the method of authority (1934, p. 236; 1986a, p. 251). It is likely that Peirce's usage of the terms "infallible", "fallible", and especially "fallibilism", his name for an attitude essential to the method of science, was to some extent an allusion to the enunciation of papal infallibility. The odd example discussed in Section 3.5 above ("Some woman is adored by all catholics", etc.), which Peirce used to illustrate the notion of logical consequence, is certainly an allusion to the doctrine of the Immaculate Conception (1933a, pp. 277–278).

In their respective careers, Holmes, Green, and Warner were important figures in turning American legal studies away from academic niceties to the solution of practical problems. Wiener writes,

---

[2]There are some inconsistencies between the account given of Peirce and the Metaphysical Club in (Wiener, 1949) and the one in (Fisch, 1986a; 1986b). The account of the Club in Sections 5.3.1–5.3.2 of this book follows Fisch's scholarship.

The law schools were steeped in classical syllogistic methods of applying the law to individual cases as previously decided and in the Hobbesian-Austinian view that the law was "the command of the sovereign." The Lockean view of the social contract was mingled with the Puritan idea of the Covenant with God.

...The upshot of pragmatic jurisprudence was the dissocation of the law from its scholastic accretions of eternal theological standards and imputations of original sin and hell-fire for the nonconformist and iconoclast. ...Peirce showed the influence of Green's analytical use of legal history when he pointed out, as Green had in the *American Law Review* (4 [Jan., 1870], 201), that key terms like "proximate cause" could not simply be transferred from Aristotelian physics to the laws of liability. "The idea of making the payment of considerable damages dependent on a term of Aristotelian logic or metaphysics is most shocking to any student of these subjects, and well illustrates the value of Pragmatism" (C. S. Peirce, "Proximate Cause and Effect," *Baldwin's Dictionary of Philosophy*). "Proximate cause" in civil law has to do with the negligence of a party with respect to the legal rights of others and nothing to do with spatio-temporal contiguity or a mechanical chain of causes. Rights and liabilities are determined by the civil law in the case of property damages which can even be inflicted at a distance, e.g., by hiring others to commit arson.

Green's influence on the shaping of legal pragmatism is not as well known as that of Oliver Wendell Holmes, Jr. ...The test of how good or bad a new law is becomes a matter of predicting the social consequences or public effects of enacting and enforcing the proposed law. (P. Wiener, 1973, pp. 566–567)[3]

Hence, there is an important connection between: (1) the climate of social and legal crisis, (2) the ancient meaning of *pragmaticus* as one skilled in practical affairs, especially in matters of the law, and (3) the presence of so many lawyers at the discussions of the Metaphysical Club. Peirce tried to find a single method which is exemplified in traditional practical skills but which reaches its perfect expression in mathematics and the laboratory sciences. And these highly pragmatic (in the ancient sense) men of affairs spent their valuable time in discussing Peirce's efforts.

---

[3]See (Peirce, 1902) for his definition of 'Proximate'.

## 5.3.2   Peirce develops his concept

During 1871–75, between two of his European trips, Peirce lived in Cambridge and joined in the Metaphysical Club discussions of science, philosophy, and law. It was at a meeting of the Club in late November of 1872 that Peirce gave a talk, using the word "pragmatism" in connection with his pragmatic maxim (Fisch, 1986a, pp. xxix–xxxii). This talk was later written up as the first two articles of Peirce's *Popular Science Monthly* series, described below (Fisch, 1986b, p. 29).

In the manuscript of a 1907 letter (which apparently was never sent), Peirce made a statement that sheds some light on the relation between the Metaphysical Club discussions and his own doctrine:

> To the Editor of *The Sun*
>
> Sir:—
>
> It is a well-settled rule among scientific men that every step in science, every new result, shall be credited to the name of him who first publishes it. ...The rule mentioned effectually prevents the rank and file of the scientific world from at all knowing, after a generation has elapsed, what did take place...
>
> I must count it as one of the most fortunate circumstances of a life which the study of scientific philosophy in a religious spirit has steeped in its joy, that I was able to know something of the inwardness of the early growth of several of the great ideas of the nineteenth century. By far the most interesting of these to me was the idea of pragmatism.
>
> ...Green was especially impressed with the doctrines of Bain, and impressed the rest of us with them; and finally the writer of this brought forward what we called the principle of pragmatism. Several years later, this was set forth in two articles printed in the *Popular Science Monthly* (November 1877 and January 1878) and subsequently in the *Revue Philosophique*.
>
> The particular point that had been made by Bain and that had most struck Green and through him the rest of us, was the insistence that what a man really believes is what he would be ready to act upon, and to risk much upon. The writer endeavored to weave that truth in with others which he made out for himself, so as to make a consistent doctrine of cognition. It appeared to him to be requisite to connect Bain's doctrine on one hand with physiological phenomena and on another hand with logical distinctions. (Fisch, 1986b, pp. 100–101)

Thus, Peirce admitted that he had combined his own philosophical developments with the inquiries of the Metaphysical Club (Fisch, 1986b, p. 95), and a conflict between Peirce's doctrine and the general discussions of the Club was inherent from the first.

As early as 1865, Peirce made a note about Kant's use of the word "pragmatic" in a notebook (Fisch, 1986b, p. 114). And in the manuscript of an 1865 lecture, Peirce wrote,

> ...singulars come under general terms only by accident, not by the implication of the words themselves. But the comprehension of a general term consists in the total of all possible things to which it is applicable and not merely to those which actually occur. So that singulars never can fill up this extension. 'All men', in logic, means man in general. I might perhaps enumerate all the men who have been, but I never can know that I have enumerated all who are to be. ...In short, the logical comprehension, is a total of possibles and possibles have no total of enumeration. (Peirce, 1982, p. 178)

This points to the difference between the effects which have already happened (a particular notion) and the conceivable effects (a general idea)—for example, between the consequences of a theory that have already been observed in experiments and the conceivable consequences of that theory.

In 1871 Peirce published his review of a new edition of the works of Berkeley, as mentioned in Section 4.1, which included this prototype of the pragmatic maxim:

> A better rule [than Berkeley's] for avoiding the deceits of language is this: Do things fulfil the same function practically? Then let them be signified by the same word. Do they not? Then let them be distinguished. If I have learned a formula in gibberish which in any way jogs my memory so as to enable me in each single case to act as though I had a general idea, what possible utility is there in distinguishing between such a gibberish and formula and an idea? Why use the term a general idea in such a sense as to separate things which, for all experiential purposes, are the same? (Peirce, 1958b, p. 34; 1984, p. 483)

In other words, if two ideas do not have the same effect, then let them be distinguished; but if they do have the same effect, to distinguish them is nonsense. In 1908, Peirce recalled using the word "pragmatism" in connection with this idea attributed in part to Berkeley:

> In 1871, in a Metaphysical Club in Cambridge, Massachusetts, I used to preach this principle as a sort of logical gospel, representing the unformulated method followed by Berkeley, and in

conversation about it I called it "Pragmatism." (Peirce, 1935, p. 328)[4]

And regarding his choice of the word, Peirce said in a 1905 article, 'What Pragmatism Is':

> ...For this doctrine [the writer] invented the name *pragmatism*. Some of his friends wished him to call it *practicism* or *practicalism* (perhaps on the ground that πρακτικος is better Greek than πραγματικος). But for one who had learned philosophy out of Kant, as the writer, along with nineteen out of every twenty experimentalists who have turned to philosophy, had done, and who still thought in Kantian terms most readily, *praktisch* and *pragmatisch* were as far apart as the two poles, the former belonging in a region of thought where no mind of the experimentalist type can ever make sure of solid ground under his feet, the latter expressing relation to some definite human purpose. Now quite the most striking feature of the new theory was its recognition of an inseparable connection between rational cognition and rational purpose; and that consideration it was which determined the preference for the name *pragmatism*. (Peirce, 1934, pp. 273–274)[5]

In other words, there is "an inseparable connection between" theory and practice or between an idea and its consequences.

As stated above, Peirce published a version of his original Metaphysical Club talk on pragmatism in the first two articles of an unfinished series of six printed in the *Popular Science Monthly* during 1877–1878.[6] In the first article of the series, 'The Fixation of Belief', Peirce was concerned with inquiry as the passage from a state of irritation or doubt to a state of belief. He described four methods of making this passage and thus fixing belief (1934, pp. 233–247; 1986a, pp. 248–257): The first method of tenacity, in which the individual seizes on the first notion that comes to mind, is motivated by the individual's desire merely to escape from the irritation of doubt. The second method of authority, in which the group enforces a uniform belief, is motivated by the group's desire to avoid the doubts that arise when different individuals arrive at different beliefs through the method of tenacity.

> ...When complete agreement could not otherwise be reached, a general massacre of all who have not thought in a certain

---

[4]It is likely that the name "Metaphysical Club" is an ironical allusion to Samuel Johnson's Literary Club, which was made famous in Boswell's life of Johnson (Boswell, 1952).

[5]James used "practicalism" in his 1898 lecture, quoted in Section 5.3.3 below.

[6]In *Writings*, the six *Popular Science Monthly* articles appear consecutively (1986a, pp. 242–338), but in *Collected Papers*, they must be reassembled (1934, pp. 223–271; 1932, pp. 389–432; 1935, pp. 283–301; 1932, pp. 372–388).

way has proved a very effective means of settling opinion in a country. If the power to do this be wanting, let a list of opinions be drawn up, to which no man of the least independence of thought can assent, and let the faithful be required to accept all these propositions, in order to segregate them as radically as possible from the influence of the rest of the world.

This method has, from the earliest times, been one of the chief means of upholding correct theological and political doctrines, and of preserving their universal or catholic character. In Rome, especially, it has been practised from the days of Numa Pompilius to those of Pius Nonus. This is the most perfect example in history... (Peirce, 1934, p. 236; 1986a, pp. 250–251)[7]

But this authority cannot extend to other cultures, to other times and places. The third method of *a priori*, in which the individual tries to reconcile private thoughts on these matters with the outer authority, is again motivated by the individual's desire to escape from the irritation of doubt, but these doubts arise from the group's practice of the method of authority.

This method resembles that by which conceptions of art have been brought to maturity. The most perfect example of it is to be found in the history of metaphysical philosophy. Systems of this sort have not usually rested upon any observed facts, at least not in any great degree. They have been chiefly adopted because their fundamental propositions seemed "agreeable to reason." This is an apt expression; it does not mean that which agrees with experience, but that which we find ourselves inclined to believe. (Peirce, 1934, pp. 238–239; 1986a, p. 252)

... This method is far more intellectual and respectable from the point of view of reason than either of the others which we have noticed. But its failure has been the most manifest. It makes of inquiry something similar to the development of taste; but taste, unfortunately, is always more or less a matter of fashion, and accordingly metaphysicians have never come to any fixed agreement... (1934, p. 241; 1986a, p. 253)

The fourth method of science, in which the community of investigators determines belief "by nothing human, but by some external permanency— by something upon which our thinking has no effect", is motivated by the

---

[7]The "list of opinions" is probably Peirce's allusion to Pius IX's *Syllabus*, mentioned in Section 5.3.1 above. Numa, the legendary first king of Rome, created the priestly offices and served as the first *pontifex* (Plutarch, 1914, pp. 339–347). For Peirce to imply that an unbroken line connected Numa, the first pagan pontifex, to Pius IX, the contemporary Catholic pontiff, was for him to follow Gibbon's argument that the Church had incorporated into itself pagan practices, ceremonies, etc. (Gibbon, 1909, pp. 198–227)

desire to escape the doubt that our beliefs may have been "determined by any circumstance extraneous to the facts." (1934, p. 242; 1986a, p. 253) The method of science does not cause new doubts through its practice; it is different from the first three methods both because it involves the concept of a reality that does not depend on our thinking and because one can be mistaken in its practice. Because there is no reality that the first three methods seek to discover, *according to the methods themselves* there can be no mistake made in their practice. But for the same reason, the first three methods are destined to be plagued by endless disputes over the doubts that necessarily arise from their practice (1934, pp. 244–245; 1986a, pp. 254–255).

In the second article, 'How to Make Our Ideas Clear', Peirce further criticized the four methods of inquiry, and especially the *a priori* method as used by Descartes and Leibniz (1934, pp. 249–255; 1986a, pp. 258–263). Neither doubt nor belief are simple voluntary actions, and imaginary doubt cannot serve as the start of real inquiry. In this second article, the pragmatic maxim (quoted at the beginning of this chapter) was given as the definition of the third grade of clearness—familiarity being the first and abstract definition the second—as it applies to our ideas or conceptions:

> Consider what effects, which might conceivably have practical bearings, we conceive the object of our conception to have. Then, our conception of these effects is the whole of our conception of the object. (Peirce, 1934, p. 258; 1986a, p. 266)

But the word "pragmatism" does not appear in this maxim, nor does it appear anywhere else in the article. Peirce followed the statement of his maxim with three examples of concepts to which the maxim applies: hardness, from mineralogy; weight, and therefore force, from physics; and reality, from the method of science, part of which was quoted and discussed in Section 4.1 above. The analysis of the concept of reality was especially important, because it is the basic concept underlying the method of science and distinguishing that method from the other three (1934, pp. 265–270; 1986a, pp. 271–275).

In the third and fourth papers of the series Peirce attempted to reach the third grade of clearness regarding the concept of probability. The fifth and sixth papers are concerned with the concept of nature and the modes of inference in science, respectively.

### 5.3.3 James popularizes the word

Peirce's usage of the word "pragmatism" had its public debut when James made the following reference to it in his lecture, 'Philosophical Conceptions and Practical Results', given at the University of California, Berkeley, in August 1898:

I will seek to define with you merely what seems to be the most likely direction in which to start upon the trail of truth. Years ago this direction was given to me by an American philosopher whose home is in the East, and whose published works, few as they are and scattered in periodicals, are no fit expression of his powers. I refer to Mr. Charles S. Peirce, with whose very existence as a philosopher I dare say many of you are unacquainted. He is one of the most original of contemporary thinkers; and the principle of practicalism—or pragmatism, as he called it, when I first heard him enunciate it at Cambridge in the early '70's—is the clue or compass by following which I find myself more and more confirmed in believing we may keep our feet upon the proper trail.

Peirce's principle, as we may call it, may be expressed in a variety of ways, all of them very simple. In the *Popular Science Monthly* for January, 1878, he introduces it as follows: The soul and meaning of thought, he says, can never be made to direct itself towards anything but the production of belief, belief being the demicadence which closes a musical phrase in the symphony of our intellectual life. Thought in movement has thus for its only possible motive the attainment of thought at rest. But when our thought about an object has found its rest in belief, then our action on the subject can firmly and safely begin. Beliefs, in short, are really rules for action; and the whole function of thinking is but one step in the production of habits of action. ... To attain perfect clearness in our thoughts of an object, then, we need only consider what effects of a conceivably practical kind the object may involve—what sensations we are to expect from it, and what reactions we must prepare. Our conception of these effects, then, is for us the whole of our conception of the object, so far as that conception has any positive significance at all. (James, 1968, p. 348)

Note that James was paraphrasing the maxim, distorting it while at the same time implying that he was actually quoting it: He substituted *effects* (particular consequences) for Peirce's *conceivable effects* ("bearing upon conduct"). To continue James's lecture,

This is the principle of Peirce, the principle of pragmatism. I think myself that it should be expressed more broadly than Mr. Peirce expresses it. The ultimate test for us of what a truth means is indeed the conduct it dictates or inspires. But it inspires that conduct because it first foretells some particular turn to our experience which shall call for just that conduct from us. And I should prefer for our purposes this evening to

> express Peirce's principle by saying that the effective meaning
> of any philosophic proposition can always be brought down to
> some particular consequence, in our future practical experience,
> whether active or passive; the point lying rather in the fact
> that the experience must be particular, than in the fact that it
> must be active. (James, 1968, pp. 348–349)

For Peirce, it is not the actual effects (which have not yet happened) that
have bearing upon our conduct (which also has not yet happened); rather,
it is our *beliefs* regarding conceivable effects that have such bearing on our
conduct.

In late 1904 or early 1905, Peirce wrote a letter to his former student,
Christine Ladd-Franklin, in which he described the difference between
his pragmatism and that of James, and she included the letter in a 1916
memorial article about him. He wrote to her,

> For in a forthcoming number of the *Monist*, I am to have an
> article about pragmatism, explaining what I conceive it to be.
> Although James calls himself a pragmatist, and no doubt he
> derived his ideas on the subject from me, yet there is a most
> essential difference between his pragmatism and mine. My
> point is that the meaning of a *concept*... lies in the manner
> in which it could *conceivably* modify purposive action, and *in
> this alone*. James, on the contrary, ... in defining pragmatism,
> speaks of it as referring ideas to *experiences*, meaning evidently
> the sensational side of experience, while I regard *concepts* as
> affairs of habit or disposition, and of how we should react
> ... There never was the smallest disloyalty on James's part. On
> the contrary, he has dragged in mention of me whenever he
> could. (Ladd-Franklin, 1916, pp. 718–719)

It does seem odd that James did not first give "Peirce's principle" accurately
in Peirce's own words and then go on to explain the differences between
his and Peirce's pragmatism. Nevertheless, Peirce believed that James was
only trying to give him due credit. He was actually quite laudatory in
a manuscript of 1911, where he used the "snarl of twine" metaphor for
himself:

> After studying William James on the intellectual side for half a
> century—for I was not acquainted with him as a boy—I must
> testify that I believe him to be, and always to have been during
> my acquaintance with him, about as perfect a lover of truth as
> it is possible for a man to be; and I do not believe there is any
> definite limit to man's capacity for loving the truth.
>
> ... His comprehension of men to the very core was most won-
> derful. Who, for example, could be of a nature so different from

his as I? He so concrete, so living; I a mere table of contents, so abstract, a very snarl of twine. Yet in all my life I found scarce any soul that seemed to comprehend, naturally, [not] my concepts, but the mainspring of my life better than he did. He was even greater [in the] practice than in the theory of psychology. (Peirce, 1935, pp. 130–131)[8]

It seems that James must have understood the ideas discussed in the Metaphysical Club in his own way and that he did not feel the word should be restricted to Peirce's technically difficult doctrine; instead, he applied the word to his own understanding and perhaps to the collective inquiries of the Club. But it also seems that the other members of the Club thought of the word as applying only to Peirce's doctrine, since they had not used the word in their writings.

John Dewey wrote about the difference between Peirce's and James's pragmatism in his 1916 memorial essay, 'The Pragmatism of Peirce':

> Now the curious fact is that Peirce puts more emphasis upon practise (or conduct) and less upon the particular; in fact, he transfers the emphasis to the general. The following passage is worth quotation because of the definiteness with which it identifies meaning with both the future and with the general. (Dewey, 1923, p. 303)

Dewey then quoted only a part of the following passage from Peirce's 1905 article, 'What Pragmatism Is':

> The rational meaning of every proposition lies in the future. How so? The meaning of a proposition is itself a proposition. Indeed, it is no other than the very proposition of which it is the meaning: it is a translation of it. But of the myriads of forms into which a proposition may be translated, what is that one which is to be called its very meaning? It is, according to the pragmaticist, that form in which the proposition becomes applicable to human conduct, not in these or those special circumstances, nor when one entertains this or that special design, but that form which is most directly applicable to self-control under every situation, and to every purpose. This is why he locates the meaning in future time; for future conduct is the only conduct that is subject to self-control. But in order that that form of the proposition which is to be taken as its meaning should be applicable to every situation and to every purpose upon which the proposition has any bearing, it must be simply the general description of all the experimental

---

[8]The bracketed words were supplied by Hartshorne and Weiss, the editors.

> phenomena which the assertion of the proposition virtually predicts. (Peirce, 1934, pp. 284–285)

Then Dewey continued,

> Or, paraphrasing, pragmatism identifies meaning with forma-
> tion of a habit, or way of acting having the greatest generality
> possible, or the widest range of application to particulars. Since
> habits or ways of acting are just as real as particulars, it is
> committed to a belief in the reality of "universals". (Dewey,
> 1923, p. 303)

In Dewey's later essay, 'The Development of American Pragmatism', he had this to add:

> ...In one sense, one can say that [James] enlarged the bearing
> of the principle by the substitution of particular consequences
> for the general rule or method applicable to future experience.
> But in another sense this substitution limited the application
> of the principle, since it destroyed the importance attached by
> Peirce to the greatest possible application of the rule, or the
> habit of conduct—its extension to universality. That is to say,
> William James was much more of a nominalist than Peirce.
> (Dewey, 1963, p. 18)

James helped to arrange two lecture series for Peirce in Cambridge: the 1898 Cambridge Conference lectures, which Peirce called 'Reasoning and the Logic of Things', and the Lowell Lectures of 1903, in which Peirce gave an account of his own version of pragmatism. Peirce mentioned the 1903 lectures[9] in the letter to Ladd-Franklin:

> In the spring of 1903 I was invited, by the influence of James,
> Royce, and Münsterberg, to give a course of lectures in Harvard
> University on Pragmatism. I had intended to print them;
> but James said he could not understand them himself and
> could not recommend their being printed. I do not myself
> think there is any difficulty in understanding them, but all
> modern psychologists are so soaked with sensationalism that
> they can not understand anything that does not mean that,
> and mistranslate into the ideas of Wundt whatever one says
> about logic. (Ladd-Franklin, 1916, pp. 719–720)

One can only try to imagine how Peirce began to feel about the impending dissociation of his concept from the word "pragmatism". He had intended to use his concept to establish metaphysics by the method of science. As the word was lost to the ensuing academic philosophical disputes over its

---

[9]See (Peirce, 1934, pp. 13–131) for manuscripts of the 1903 lectures.

meaning, disputes which were a throwback to the *a priori* method, Peirce must have found it excruciating to think that this was the very method which would ensure that the disputes would be endless and that nothing would be contributed to science or to a scientific metaphysics. Perhaps such feelings of imminent doom led Peirce to pursue once more his desire to give proofs in scientific metaphysics by means of a mathematical logic.

### 5.3.4   Peirce reacts to the ensuing confusion

We shall see in this section how Peirce coined the new word, "pragmaticism", for his own concept. In a 1905 letter to the Italian philosopher, Mario Calderoni, Peirce objected to the confusion between his definition of pragmatism and that of James:

> ...I deny that pragmaticism as originally defined by me made the intellectual purport of symbols to consist in our conduct. On the contrary, I was most careful to say that it consists in our *concept* of what our conduct *would* be upon *conceivable* occasions. (Peirce, 1958b, p. 166)

> ...Man seems to himself to have some glimmer of co-understanding with God, or with Nature. The fact that he has been able in some degree to predict how Nature will act, to formulate general "laws" to which future events conform, seems to furnish inductive proof that man really penetrates in some measure the ideas that govern creation. Now man cannot believe that creation has not some ideal purpose. If so, it is not mere action, but the development of an idea which is the purpose of thought; and so a doubt is cast upon the ultra pragmatic notion that action is the *sole* end and purpose of thought. (pp. 169–170)

In the same letter, Peirce wrote, "The truth of pragmaticism may be proved in various ways. I would conduct the argument somewhat as follows." (1958b, p. 167)[10] Then he went on to sketch an argument which followed the sequence of topics in the 1903 lectures. As mentioned in Section 4.1 above, Peirce had suggested in his 1871 review of Berkeley that he then intended or proposed to use the logic of relations to give a proof of realism, and he had made a similar claim in the 1898 lectures. Such a proof would have been an example of using exact or mathematical logic to solve a problem in metaphysics, which would have been in accord with his chart of the organization of mathematics, logic, philosophy, and the sciences discussed in Section 4.2. By the time Peirce wrote the 1905 letter,

---

[10]Only the first half of the letter appears in the place cited; the second half was published in (Fisch and Kloesel, 1982).

he may have decided that a proof of pragmaticism would supersede or include his earlier proposal to give a proof of realism.

During 1905–1906, Peirce published three articles in *The Monist* regarding his own version of pragmatism (1934, pp. 272–313; 1933b, pp. 411–463). He described them in the letter to Ladd-Franklin:

> ...my *Monist* Article (already sent in and accepted) is to explain what my position is; and I desire to follow it up by two others, of which the first shall show how this principle at once affords solutions to a great variety of problems, and shall show what the general color of those solutions is, while the third article shall show what facts and phenomena I appeal to as proving the truth of the pragmatist principle. (Ladd-Franklin, 1916, p. 718)

In the first article, 'What Pragmatism Is', Peirce explained his concept and gave it the new name, "pragmaticism" (1934, pp. 276–277). He also apparently proposed to give a "proof" of it:

> ...But [pragmaticism's] capital merit, in the writer's eyes, is that it more readily connects itself with a critical proof of its truth. Quite in accord with the logical order of investigation, it usually happens that one first forms an hypothesis that seems more and more reasonable the further one examines into it, but that only a good deal later gets crowned with an adequate proof. The present writer having had the pragmatist theory under consideration for many years longer than most of its adherents, would naturally have given more attention to the proof of it. ...In the present article there will be space only to explain just what this doctrine ...really consists in. Should the exposition be found to interest readers of *The Monist*, they would certainly be much more interested in a second article which would give some samples of the manifold applications of pragmaticism (assuming it to be true) to the solution of problems of different kinds. After that, readers might be prepared to take an interest in a proof that the doctrine is true—a proof which seems to the writer to leave no reasonable doubt on the subject, and to be the one contribution of value that he has to make to philosophy. (Peirce, 1934, pp. 277–278)

The second article, 'Issues of Pragmaticism', appeared six months after the first; Peirce began it by giving the original maxim and a restatement:

> Pragmaticism was originally enounced in the form of a maxim, as follows: Consider what effects that might *conceivably* have practical bearings you *conceive* the objects of your *conception*

to have. Then your *conception* of those effects is the whole of
your *conception* of the object.

I will restate this in other words, since ofttimes one can thus
eliminate some unsuspected source of perplexity to the reader.
This time it shall be in the indicative mood, as follows: The
entire intellectual purport of any symbol consists in the total
of all general modes of rational conduct which, conditionally
upon all the possible different circumstances and desires, would
ensue upon the acceptance of the symbol. (Peirce, 1934, p. 293)

Peirce used the third article, 'Prolegomena to an Apology for Pragmaticism',
which appeared a year after the second, to introduce his existential graphs.
These diagrams or graphs included systems for propositional, first-order,
and modal logic. The third article ended with this tantalizing remark: "In
my next paper, the utility of this diagrammatization of thought in the
discussion of the truth of Pragmaticism shall be made to appear." (1933b,
p. 463) Thus, Peirce gave his concept in the first article, he discussed
certain conceivable effects of that concept in the second article, and he
proposed to produce a real effect in the third and fourth articles—a "proof"
that pragmaticism (and therefore realism) is valid. But no fourth article
appeared (1933b, p. 427n), and it remains uncertain what part the graphs
were to have played in a full "proof" (Fisch, 1986b, pp. 362–375).

In 'What Pragmatism Is', Peirce gave this response to the increasing
notoriety of the word "pragmatism":

> ... [the writer] framed the theory that a *conception*, that is, the
> rational purport of a word or other expression, lies exclusively
> in its conceivable bearing upon the conduct of life; so that
> since obviously nothing that might not result from experiment
> can have any direct bearing upon conduct, if one can define
> accurately all the conceivable experimental phenomena which
> the affirmation or denial of a concept could imply, one will
> have therein a complete definition of the concept, and *there is
> absolutely nothing more in it.* For this doctrine he invented the
> name *pragmatism.* (Peirce, 1934, pp. 273–274)

> ... His word "pragmatism" has gained general recognition in
> a generalized sense that seems to argue power of growth and
> vitality. The famed psychologist, James, first took it up, seeing
> that his "radical empiricism" substantially answered to the
> writer's definition of pragmatism, albeit with a certain difference
> in the point of view. ... So far all went happily. But at present,
> the word begins to be met with occasionally in the literary
> journals, where it gets abused in the merciless way that words
> have to expect when they fall into literary clutches. ... So then,
> the writer, finding his bantling "pragmatism" so promoted,

feels that it is time to kiss his child good-by and relinquish it to its higher destiny; while to serve the precise purpose of expressing the original definition, he begs to announce the birth of the word "pragmaticism," which is ugly enough to be safe from kidnappers.

Much as the writer has gained from the perusal of what other pragmatists have written, he still thinks there is a decisive advantage in his original conception of the doctrine. ... At any rate, in endeavoring to explain pragmatism, he may be excused for confining himself to that form of it that he knows best. In the present article there will be space only to explain just what this doctrine (which, in such hands as it has now fallen into, may probably play a prominent part in the philosophical discussions of the next coming years), really consists in. (pp. 276–277)

... Let us now hasten to the exposition of pragmaticism itself. Here it will be convenient to imagine that somebody to whom the doctrine is new, but of rather preternatural perspicacity, asks questions of a pragmaticist. Everything that might give a dramatic illusion must be stripped off, so that the result will be a sort of cross between a dialogue and a catechism...

*Questioner:* I am astounded at your definition of your pragmatism, because only last year I was assured by a person above all suspicion of warping the truth—himself a pragmatist—that your doctrine precisely was "that a conception is to be tested by its practical effects." You must surely, then, have entirely changed your definition very recently.

*Pragmatist:* If you will turn to Vols. VI and VII of the *Revue Philosophique*, or to the *Popular Science Monthly* for November 1877 and January 1878, you will be able to judge for yourself whether the interpretation you mention was not then clearly excluded. The exact wording of the English enunciation, (changing only the first person into the second), was: "Consider what effects that might conceivably have practical bearing you conceive the object of your conception to have. Then your conception of those effects is the WHOLE of your conception of the object."

*Questioner:* Well, what reason have you for asserting that this is so?

*Pragmatist:* That is what I specially desire to tell you. But the question had better be postponed until you clearly understand what those reasons profess to prove.

*Questioner:* What, then, is the *raison d'être* of the doctrine? What advantage is expected from it?

> *Pragmatist:* It will serve to show that almost every proposition
> of ontological metaphysics is either meaningless gibberish—one
> word being defined by other words, and they by still others,
> without any real conception ever being reached—or else is
> downright absurd; so that all such rubbish being swept away,
> what will remain of philosophy will be a series of problems
> capable of investigation by the observational methods of the
> true sciences—the truth about which can be reached without
> those interminable misunderstandings and disputes which have
> made the highest of the positive sciences a mere amusement for
> idle intellects, a sort of chess—idle pleasure its purpose, and
> reading out of a book its method. In this regard, pragmati-
> cism is a species of prope-positivism. But what distinguishes
> it from other species is, first, its retention of a purified phi-
> losophy; secondly, its full acceptance of the main body of our
> instinctive beliefs; and thirdly, its strenuous insistence upon
> the truth of scholastic realism... So, instead of merely jeering
> at metaphysics, like other prope-positivists, whether by long
> drawn-out parodies or otherwise, the pragmaticist extracts
> from it a precious essence, which will serve to give life and
> light to cosmology and physics. At the same time, the moral
> applications of the doctrine are positive and potent; and there
> are many other uses of it not easily classed. (pp. 281–282)

Thus, in order to preserve the integrity of his own doctrine, Peirce gave up
the original word to the "literary journals". Because the literary philosophers know not the method of science, they spin out endless metaphysical
disputes according to their practice of the *a priori* method. There can
never be an objective means of settling such disputes, because their method
never leads to experiments, mental or physical, and because their beliefs
depend only on their own thoughts.

By 1906, the situation had become quite confusing—even radical politicians on the Continent had appropriated the word "pragmatism" to describe
their revolutionary social agendas. Wiener cites a kind of ludicrous limit
to this confusion, as reached by one G. Papini:

> The opening paragraph of G. Papini's work *Pragmatismo (1905–*
> *1911)*, a collection of his articles introducing that doctrine to
> Italian philosophers, reads: "Pragmatism cannot be defined.
> Whoever gives a definition of Pragmatism in a few words would
> be doing the most antipragmatic thing imaginable" ...Papini
> was (in 1906) echoing William James's romantic aversion to
> fixed definitions, and even mistakenly placed Peirce in the same
> boat with James...(P. Wiener, 1973, p. 552)

In a paper of 1908, Peirce mentioned this anti-definition of Papini:

Of course, the doctrine [of pragmatism] attracted no particular
attention [in 1878], for, as I had remarked in my opening
sentence, very few people care for logic. But in 1897 Professor
James remodelled the matter, and transmogrified it into a
doctrine of philosophy, some parts of which I highly approved,
while other and more prominent parts I regarded, and still
regard, as opposed to sound logic. About the time Professor
Papini discovered, to the delight of the Pragmatist school, that
this doctrine was incapable of definition, which would certainly
seem to distinguish it from every other doctrine in whatever
branch of science, I was coming to the conclusion that my poor
little maxim should be called by another name; and accordingly,
in April, 1905 I renamed it *Pragmaticism*. I had never before
dignified it by any name in print, except that, at Professor
Baldwin's request, I wrote a definition of it for his *Dictionary
of Psychology and Philosophy*. I did not insert the word in the
*Century Dictionary*, though I had charge of the philosophical
definitions of that work... (Peirce, 1935, p. 329)[11]

Also in 1908, Arthur O. Lovejoy, a student of James at Harvard, published
a two part article, 'The Thirteen Pragmatisms'. Although Lovejoy's
article does not say very much about Peirce, its title alone suggests that
"pragmatism" had acquired a powerful appeal, since so many and so varied
individuals had already appropriated the word.

### 5.3.5 After Peirce's death

Peirce died destitute, in 1914. In 1916, *The Journal of Philosophy, Psychology, and Scientific Methods* published a Peirce memorial issue which
included the memoir by Ladd-Franklin (1916) and Dewey's memorial article, 'The Pragmatism of Peirce' (1923), both quoted above. In 1923, there
was sufficient interest in Peirce's philosophy that Morris Cohen edited the
series of six articles on scientific method from the *Popular Science Monthly*
of 1877–78,[12] a series of five philosophical articles from *The Monist* of 1891–
93,[13] and Dewey's memorial article, giving the collection the title, *Chance,
Love, and Logic*. This was the first book publication of any philosophical
writing by Peirce. (M.R. Cohen, 1923)

In 'What Pragmatism Is', Peirce had called for an ethics of terminology
which would respect the definition given to a technical term by the one
who introduced it:

---

[11]See (Peirce, 1934, pp. 1–3) for his definition of 'Pragmatic and Pragmatism' from
Baldwin's *Dictionary*, which was quoted in part at the beginning of Section 5.1.

[12]See footnote 6 on page 92.

[13]See (Peirce, 1935, pp. 11–45, 86–113, 155–177, 190–215).

Concerning the matter of philosophical nomenclature, there are a few plain considerations, which the writer has for many years longed to submit to the deliberate judgment of those few fellow-students of philosophy, who deplore the present state of that study, and who are intent upon rescuing it therefrom and bringing it to a condition like that of the natural sciences, where investigators, instead of contemning [*sic*] each the work of most of the others as misdirected from beginning to end, coöperate, stand upon one another's shoulders, and multiply incontestible results... To those students, it is submitted that no study can become scientific in the sense described, until it provides itself with a suitable technical nomenclature, whose every term has a single definite meaning universally accepted among students of the subject... It is submitted that the experience of those sciences which have conquered the greatest difficulties of terminology, which are unquestionably the taxonomic sciences, chemistry, mineralogy, botany, zoölogy, has conclusively shown that the one only way in which the requisite unanimity and requisite ruptures with individual habits and preferences can be brought about is so to shape the canons of terminology that they shall gain the support of *moral principle* and of every man's sense of decency; and that, in particular (under defined restrictions), the general feeling shall be that he who introduces a new conception into philosophy is under an obligation to invent acceptable terms to express it, and that when he has done so, the duty of his fellow-students is to accept those terms, and to resent any wresting of them from their original meanings, as not only a gross discourtesy to him to whom philosophy was indebted for each conception, but also as an injury to philosophy itself... (Peirce, 1934, pp. 274–275)

In a printed supplement to the 1903 lectures on pragmatism, Peirce had lamented his own transgression of this "duty" in regard to De Morgan's "logic of relations":

I must, with pain and shame, confess that in my early days I showed myself so little alive to the decencies of science that I presumed to change the name of this branch of logic, a name established by its author and my master, Augustus De Morgan, to "the logic of relatives." I consider it my duty to say that this thoughtless act is a bitter reflection to me now, so that young writers may be warned not to prepare for themselves similar sources of unhappiness. I am the more sorry, because my designation has come into general use. (Peirce, 1933a, p. 367n2)

In the academic philosophical disputes over pragmatism lies the proximate cause of the loss of the traditional, common sense, positive meaning of the word "pragmatic"; certain philosophers were negligent of the rights of others in thus debasing the currency of our language.

There have, of course, been syncretic attempts to find some englobing principle or principles that would unify (or at least resolve the apparent contradictions in) this Babel of usages of the word "pragmatism". From Lovejoy's 'The Thirteen Pragmatisms' (1908) on down to Thayer's comprehensive *Meaning and Action: A Critical History of Pragmatism* (1981a), we find that these more or less scholarly attempts are more or less successful in classifying the differences between the doctrines chosen as examples of pragmatism. But in the final analysis, it seems that no single coherent definition can embrace all of these doctrines. And so, from the zenith of Peirce's attempt to follow an ethics of terminology by using a word to express the exact relation between theory and practice, passing through a decline during which the controversies of literary and metaphysical philosophers constituted contributory negligence, we have reached a nadir wherein the word applies to cynical politicians and demagogues who would use statements as mere actions without regard to any real meaning.

# Chapter 6

# Peirce in the History of Mathematical Logic

In this chapter, we take up two questions which naturally arise from the material in Chapters 2 and 3: first, what is Peirce's place in the history of mathematical logic? and second, why has his work not been better recognized? We take two approaches to these questions: a dialogue in Section 6.1 and an exposition in Section 6.2.

## 6.1   Dialogue on Peirce's place

The setting for the dialogue is an imaginary International Congress on the History of Mathematics, Logic, and Philosophy. There are six characters:

(1) HISTORIAN, who exemplifies the viewpoint and positions taken in this book;

(2) LOGICIAN, who has seriously studied many of the original papers in logic—especially those of Löwenheim (1915) and Skolem (1920; 1923)—but who has not yet thought much about the entire line of development connecting Boole, De Morgan, Peirce, Schröder, Löwenheim, and Skolem;

(3) MATHEMATICIAN, who has a strong interest in the history of mathematics and some interest in logic, but who has not studied the original papers in logic;

(4) FORMALIST and (5) FREGEAN, two philosophical logicians who have studied mathematical logic and its history, but who have not studied mathematics beyond the undergraduate level; and

(6) PHILOSOPHER, who has a strong interest in Peirce and some interest in logic, but who has studied neither mathematics nor logic beyond the undergraduate level. The three days or episodes of the dialogue take place in a coffee shop following each of the three invited talks in a retrospective series on the history of mathematical logic titled, 'From Gödel to Frege'.

The dialogue involves material from all the preceding chapters of this book. Some of the points made in the Third Day depend particularly on Sections 2.1–2.2.

## 6.1.1   First Day: Gödel

(Following LOGICIAN's lecture on Gödel's completeness and incompleteness theorems; HISTORIAN, LOGICIAN, and PHILOSOPHER are seated together in the coffee shop...)

HISTORIAN: Logician, that was an excellent lecture you just gave.

PHILOSOPHER: I enjoyed it, too, although I wish that I could understand these results of Gödel's better than I do. I've tried to get something out of both Hofstadter's book (1979) and Nagel and Newman's book (1958), but I still feel that it all eludes me.

HISTORIAN: But surely that applies to your field, too, doesn't it? How successful can popularizations of Plato, Kant, or Peirce ever really be?

LOGICIAN: I've been thinking about something along these lines, myself: There exist popular accounts of incompleteness but not of completeness; this seems to me to be circumstantial evidence that the completeness theorem is much more difficult than the incompleteness theorem for non-professionals in logic to understand. Do you find that to be so, Philosopher?

PHILOSOPHER: Well, I believe that the incompleteness theorem is somehow about not being able to prove propositions that refer to themselves in a self-contradictory way—propositions such as, "This statement is false," and, "There are no absolute truths." That strikes me as both intriguing and reasonable—we need to beware of such paradoxes or 'outlaw statements' in ordinary language and, I guess, even more so in symbolical languages like mathematics and mathematical logic. But the completeness theorem never really made any sense to me. I just can't conceive of a gap between what we can prove and what is really true. It seems to me that it wouldn't really be *logic* unless what we can prove coincides *exactly* with what is really true.

HISTORIAN: If there were only one universe of discourse at stake, then the completeness theorem would have to be about that one universe. But if we are concerned with many different universes— or models—then the completeness theorem assures us that what we can prove includes what is really true in all of those universes, although it gives us nothing much that would distinguish between them.

LOGICIAN: And many mathematicians are so used to thinking of different models or universes of discourse, that it seems natural to some of them to wonder whether a certain formal system of rules of deduction in a certain formal language could succeed in making it possible to deduce all of the universally valid sentences.

HISTORIAN: And I confess that I too have been thinking "along these lines", as you just now put it, Logician—in fact, about two lines of development in the history of mathematical logic which come together in Gödel's completeness proof. I call them the deduction-theoretic line and the model-theoretic line; each of them has a distinct history that can be traced back into the nineteenth century.

LOGICIAN: As I understand it, the completeness conjecture was first made in the 1928 edition of Hilbert and Ackermann (Dreben and van Heijenoort, 1986, pp. 47–48), even though Gödel wrote at the start of his completeness paper, "when such a procedure is followed [that is, to derive mathematics or arithmetic formally] the question at once arises whether the initially postulated system of axioms and principles of inference is complete." (Gödel, 1986, p. 103; Dreben and van Heijenoort, p. 44)

HISTORIAN: Yes, and the conjecture didn't appear in Hilbert and Ackermann's 1938 edition, since the *result* could be given instead.

LOGICIAN: You know, Gödel's statement has always struck me a bit strangely. I can't imagine him as being arrogant in any way, but whatever could he have meant by saying that "the question at once arises"?

HISTORIAN: Indeed, it didn't arise for Russell and Whitehead some fifteen years or twenty years earlier, nor did it arise for Frege some forty or fifty years earlier—and they were certainly following "such a procedure". I think it has to do with what we were just now saying to our friend, Philosopher. You were saying that it is natural for mathematicians who work with different structures—in abstract algebra or geometry, for example—at least to comprehend the question whether the deductive logic they use actually suffices to derive all of the universally valid propositions. And I was implying that if one works mainly with a single universe of discourse in which to derive all of arithmetic or mathematics—the universe of sets or of set theory and type theory, for example—the completeness conjecture would not seem to be so real.

LOGICIAN: So the question could only "at once" arise for someone who is thinking both syntactically and semantically, and I expect that this has to do with your two lines of development. But this reminds me of something I meant to ask you about: I remarked in my lecture that Gödel acknowledged his use of Skolem normal form in the completeness proof (1986, p. 109n14), but didn't Gödel make some remarks about why Skolem himself didn't prove the completeness theorem?

HISTORIAN: Yes, in two letters he wrote to Hao Wang that were printed at the beginning of *From Mathematics to Philosophy* (Wang, 1974, pp. 8–11; Dreben and van Heijenoort, 1986, pp. 51–52). The gist of it is: first, Gödel said that the completeness theorem is an almost trivial consequence of Skolem's 1923 paper, and second, Gödel claimed that Skolem actually stated the completeness theorem in 1928, but that he failed to prove it because he had the wrong philosophical outlook—that is to say, the wrong epistemological attitude toward metamathematics and non-finitary reasoning.

LOGICIAN: Well...

HISTORIAN: I believe that Gödel was misreading Skolem's use of "contradiction", "consistent proposition", and "derivable" as referring to syntax (Skolem, 1977c, p. 519), whereas Skolem actually meant the terms semantically; that is to say, Skolem meant "unsatisfiable", "satisfiable", and "logical consequence". In my opinion, Skolem was not really stating a completeness theorem, and van Heijenoort makes this point in his introduction to Skolem's 1920 paper (van Heijenoort, 1977a, pp. 252–253).

LOGICIAN: Then one might say it wasn't so much that Skolem had the wrong outlook to prove the theorem as that he had the wrong outlook to state it?

HISTORIAN: Of course, it would be tendentious to say that Skolem's outlook was "wrong", but he did have an outlook in which the syntactical aspect was informal—that is to say, in the sense that most mathematicians use an informal syntax. Maybe it would be better to say that he did not have the necessary *interest* in formal deduction to conjecture the completeness theorem.

PHILOSOPHER: What did Skolem prove?

LOGICIAN: The Löwenheim-Skolem Theorem says that if a first-order theory has a model it has a countable model. In particular, applying it to set theory, there is a countable model of set theory, even though within set theory it is possible to prove that an uncountable set exists. In other words, there is a countable model of set theory in which "uncountable" does not *really* mean uncountable.

PHILOSOPHER: Well, Historian, the question of completeness may "at once" arise for you and Logician, but I can catch only a glimpse of its profundity. I'm going to have to do some last minute work on a lecture I'm giving tomorrow about Peirce's philosophy of pragmatism. I regret that the time of my lecture conflicts with the lecture on Russell and Whitehead, but I'll have to see you two the day after tomorrow, at the Frege lecture.

LOGICIAN: Forgive us, Philosopher. We really didn't mean to leave you out of the conversation. I hope your lecture goes well.

HISTORIAN: So do I. And we'll try to 'complete' your understanding later, because I really do want to return to the question of whether it is easier for non-professionals to understand the incompleteness theorem than the completeness theorem.

(PHILOSOPHER leaves.)

HISTORIAN: Logician, in your lecture you also reminded us that Gödel, in his completeness proof used the technique of building partial models which both Löwenheim and Skolem had used earlier. I'm certain you agree with me that the Löwenheim-Skolem Theorem is important, but my considerations of the model-theoretic line of development have led me to investigate just where it came from. Have you thought very much about this?

LOGICIAN: Well, there are the two papers of Skolem from 1920 and 1923.

HISTORIAN: But where did these two papers come from?

LOGICIAN: I have often wondered about this. Of course, Skolem mentions Löwenheim's 1915 paper. But there is some peculiar notation and terminology in all three of these papers. Take Skolem's 1920 paper: Instead of writing P(i,j,...) he writes $P_{ij...}$; he calls these relative coefficients. He uses $\sum$ and $\prod$ for the quantifiers. This agrees with Löwenheim's paper, but Löwenheim's notation and terminology are even further away from ours. Furthermore, Löwenheim uses equations rather than stand-alone propositions. So Löwenheim and, to a lesser extent, Skolem seem to belong to a subculture of some kind—one feels that one has run into a tradition different from that which one is used to.

HISTORIAN: Since Skolem's two papers are to an important extent based on Löwenheim's 1915 paper, the crucial question is, "Where did Löwenheim's paper come from?"

LOGICIAN: Yes, indeed, that is the crucial question. There is something of a mystery here. I think the story is as follows: Löwenheim first learned his logic from Schröder's three volumes. He thought that $\sum$ and $\prod$ in Schröder denoted "sums" and "products", but Schröder's work is confusing: Who can tell what these infinite sums and products were supposed to be? Thus Löwenheim was living in a kind of logical darkness. Then he read *Principia Mathematica* and a great light dawned on him. He saw that Schröder's system could be translated into genuine quantificational logic— or more precisely, that certain formulas of *Principia* could be translated into Schröder's system. But since he had been brought up on Schröder's notation, he kept using it out of habit or out of deference. Nevertheless, from that time on he thought within the framework of *Principia*, even though he used the notation and terminology of Schröder.

HISTORIAN: That is more or less what I used to think before I looked into the matter. But there is a certain emphasis on elevating *Principia* and derogating Schröder that I no longer think is correct. In fact, it is sort of curious in itself that to diminish Schröder's importance is such a common attitude, when so few logicians really know the material—it's just not sound science or history of science, is it?

LOGICIAN: Well, now that you put it that way, I confess that I have never been interested in looking at Schröder's work, and I guess that was a result of the same attitude you've just described.

HISTORIAN: I must go now, but let's continue this discussion tomorrow.

## 6.1.2 Second Day: Russell and Whitehead

(Following FORMALIST's lecture on *Principia Mathematica*, by Russell and Whitehead; LOGICIAN and PHILOSOPHER are seated together in the coffee shop...)

LOGICIAN: You know, Historian, I've been thinking about the conversation we had yesterday, and I am becoming more and more intrigued by the possible circumstances under which the Löwenheim-Skolem Theorem arose.

HISTORIAN: That was one of the two biggest clues in my 'case' of the model-theoretic line of development in logic. The other was. . .

LOGICIAN: Wait! Here comes our lecturer.

(FORMALIST joins them.)

LOGICIAN: Professor Formalist, I enjoyed your lecture very much. Professor Historian and I have been talking about the Löwenheim-Skolem Theorem, and we've been particularly concerned both with the influence of Löwenheim's 1915 paper on Skolem and with just how the ideas in Löwenheim's paper arose.

FORMALIST: I don't think that Löwenheim's paper made much of a contribution to modern logic.

LOGICIAN: Let's see, now. . . Would you at least grant that the statement of the theorem together with the method of proof—that is, the construction of the partial models—are two points of contribution?

(FORMALIST nods.)

LOGICIAN: Of course, Skolem chose not to use the partial models in his 1920 paper, relying on the axiom of choice instead. But since Skolem wished to avoid the set-theoretic implications of the axiom of choice in his 1923 paper, he returned to the partial models. (Dreben and van Heijenoort, 1986, p. 52)

FORMALIST: But there were gaps in Löwenheim's proof of his theorem (van Heijenoort, 1977a, pp. 231, 252–253, 290); moreover, the paper was written in the 'algebra of logic' tradition of Peirce and Schröder, and thus it suffered from the confusions and lack of clarity of that tradition. Now that I think about it, it is rather remarkable that Löwenheim was even able to arrive at a statement of his theorem—or at least at something that Skolem was able to convert into what is now known as Löwenheim's Theorem.

LOGICIAN: Historian, what's this about Peirce and Schröder?

HISTORIAN: Logician, that's what I was just about to mention earlier. Professor Formalist, I am reminded of the passage in Sherlock Holmes's adventure of 'Silver Blaze' about "the curious incident of the dog in the night". Do you recall it?

FORMALIST: Yes, yes, of course: "The dog did nothing in the night."

HISTORIAN: And, "That was the curious incident." (Doyle, 1967, p. 277) Well, I wish to call your attention to the curious incident of the first-order logic in *Principia Mathematica*—and, for that matter, in Frege's *Begriffsschrift*.

FORMALIST: There is no particular distinction between first- and second-order logic—or any other order, for that matter—neither in *Prin-*

*cipia* nor in *Begriffsschrift*.

HISTORIAN: But isn't that a *curious incident*? Even if *Principia* were essential to Löwenheim's and Skolem's papers, it wasn't sufficient, was it?

(MATHEMATICIAN arrives.)

MATHEMATICIAN: Do I detect some controversy here?

HISTORIAN: Inquiry—historical inquiry and deduction, my dear Mathematician.

LOGICIAN: I've been discussing with Historian the importance and influence of the ideas in Leopold Löwenheim's 1915 paper. Professor Formalist, however, doesn't seem to believe that the paper was very significant.

MATHEMATICIAN: I guess I think of Löwenheim as an anomaly in the development of logic from *Principia* to Gödel's completeness and incompleteness theorems.

FORMALIST: Well put! Alright, Professor Historian, I will admit that the concept of first-order logic is pretty clear in Skolem's paper of 1920 and very clear in his paper of 1923. But once the logic of *Principia* had been developed, anyone might have separated out first-order logic for special consideration.

HISTORIAN: But it was not just "anyone" who distinguished first- and second-order logic. Peirce did it in his paper of 1885, giving a general picture that Schröder preserved in his own three volumes.

LOGICIAN: Historian, earlier you were beginning to explain something about the model-theoretic line of development, and I am quite interested in the history of model theory.

MATHEMATICIAN: Well, the informal concept of a model was developing all through the nineteenth century. The consistency of non-Euclidean geometries was demonstrated by giving models, and Hilbert gave a model of his axioms of geometry in his 1899 book, *Foundations of Geometry* (Hilbert, 1971).

FORMALIST: Yes, perhaps Skolem got the idea of first-order logic from the Hilbert school, or even from Hilbert's analysis of the axioms of geometry in his 1899 book.

HISTORIAN: No, I believe Skolem's papers were based on Löwenheim's paper. Löwenheim's paper includes clear definitions of first- and second-order formulas, the notion of the logical consequence of a finite set of formulas, and set-theoretic semantics in the construction of the partial models (at least in the case of countable domains). Löwenheim's paper of 1915 and Skolem's papers of 1920 and 1923 were purely model-theoretic in a modern sense, and both of them were working with the ideas and notation of Peirce and Schröder.

FORMALIST: *Really*, Professor Historian, these anachronisms are hopelessly, even ludicrously farfetched. Face it! Peirce and Schröder's algebra of logic was a dead end.

HISTORIAN: What do you mean by "Peirce and Schröder's algebra of logic"?

FORMALIST: Boolean algebra... and Peirce had an algebra of relations.

HISTORIAN: But Professor Formalist, you are being anachronistic by imposing the present-day, much narrower sense of "algebra of logic" onto Peirce's and Schröder's work without taking into account what they actually *did*. Many of their concerns belong to the present-day field of model theory. Peirce and Schröder certainly had the quantifiers and set-theoretic semantics. Löwenheim and Skolem both wrote their early papers on Schröder's work. In the continuous line of development of ideas from Peirce to Skolem there cannot be found any "dead end", as you just put it. And by the way, I think that you just might be very interested in Norbert Wiener's Ph.D. Thesis—are you familiar with it?

FORMALIST: No.

HISTORIAN: He made a comparison between *Principia Mathematica* and Schröder's work (N. Wiener, 1953, pp. 171–175). The correspondences he found between many of Schröder's and *Principia*'s relational formulas are, to say the least, quite remarkable (N. Wiener, 1913, pp. 58–61)[1]. Perhaps Schröder's work was more important and influential than you realize.

FORMALIST: I doubt it.

HISTORIAN: Well, I expect that you *are* familiar with Russell's two essays, 'Mathematics and the Metaphysicians' and 'Logic as the Essence of Philosophy', but I would bet that you do not know that the definition of mathematics given at the beginning of the first essay (1971, pp. 59–60) is almost the same as the one given much earlier by Peirce's father, Benjamin,[2] nor that the exposition of hypothetical propositions and polyadic predicates in the second essay (1926, pp. 42–69) is virtually a paraphrase of Peirce's treatment in several of his publications.[3]

FORMALIST: I am afraid that I don't have any more time for such disputatious chitchat. I shall have to bid you all a Good Day.

(FORMALIST leaves.)

MATHEMATICIAN: What was all that about?

HISTORIAN: I guess that one can view the history of logic either as the development and influence of important ideas or as a kind of combat over personalities. Peirce often lamented the prevalence of the latter attitude in philosophy and logic. I only wanted Formalist to tell us what he thinks about the ideas, but he seems to have taken it as some kind of challenge.

LOGICIAN: Well, Historian, I assure you that *I am* interested in the ideas, and I hope that we can continue to talk about your idea of the model-theoretic line of development tomorrow.

---

[1]Wiener said in his autobiography that the Thesis was typed (N. Wiener, 1953, p. 173), but the pages cited from the Harvard library copy were handwritten.

[2]As was discussed in Section 4.2.2 of this book.

[3]For example, see (Peirce, 1933a, pp. 33–34, 38–39, 262–263, 293–295; 1984, pp. 364–366, 370).

### 6.1.3   Third Day: Frege

(Immediately following Professor FREGEAN's lecture on Frege's *Begriffss-chrift*; HISTORIAN, LOGICIAN, and PHILOSOPHER are seated together in the coffee shop. . . )

LOGICIAN: Historian, our friend Mathematician was saying yesterday that the informal notion of a model was "in the air", so to speak, during the nineteenth century. Can you say something more definite about the origin of the concept of a model?

HISTORIAN: Not anything quite so striking as that metaphor we just heard in the lecture—the one about quantification theory springing fully-formed from the head of Frege the way Athena sprang from the head of Zeus.[4] But anyway, I believe that the first definition of a model can be found in De Morgan's 1846 paper on logic, where he introduced the technical term "universe" to denote the domain of objects to which a proposition or a name applies, with all the rules of logic remaining the same in every universe (1966, p. 2). He did this in order to make the complement operation coherent by avoiding a grandiose single universe of all possible conceptions.

PHILOSOPHER: That reminds me of the term "universe of discourse".

HISTORIAN: Yes, that was Boole's term, which he used in his 1854 book (p. 42); he and De Morgan were very close colleagues. Peirce said that he had combined De Morgan's work on relations with Boole's work on an algebraic calculus of classes to get his own logic of relatives (1933a, p. 27–28; 1984, pp. 359–360), and he acknowledged De Morgan's definition of the term "universe" (1933a, p. 35; 1984, p. 366). Schröder's work is replete with references to Peirce; in fact, Kenneth Ketner gives a complete list of such passages in his bibliography of Peirce's published works (1986, pp. 89, 92, 104, 109). And Löwenheim and Skolem both did their early work on Schröder's algebra of logic (Thiel, 1977, pp. 237–238; Moore, 1988, p. 120; Fenstad, 1970, p. 9); in fact, Skolem actually wrote his bachelor's thesis on it (Moore, 1988, p. 122). Thus the model-theoretic line of development—from Boole and De Morgan through Peirce and Schröder to Löwenheim and Skolem—can be objectively established through these direct influences. In 1936, Alfred Tarski gave the accepted, rigorous definition of logical consequence—a fundamental concept of model theory (1983, p. 417)—and this marked an epoch in the development of these ideas.

And now, Logician and Philosopher, I think that we can make another approach to the question Logician raised two days ago: "Why does the incompleteness theorem seem to be easier than the completeness theorem for non-professionals in logic to understand?" I believe that it has to do with the difference between deduction theory and model theory. But Logician, please tell us what you have already thought about it.

---

[4]This locution was a favorite of Willard Quine's. A somewhat qualified version of it occurs at (Quine, 1985, p. 768). See footnote 7 on page 129, below.

LOGICIAN: Forgive me, Historian, but here come Fregean and Formalist, and I want to talk a little about the history of logic with them. But I *promise* you that we will get back to that question I raised.

(FREGEAN and FORMALIST join them.)

PHILOSOPHER: I'm very sorry I had to miss your lecture yesterday, Professor Formalist. I heard it was quite good.

FORMALIST: I thought it went well. It will be published in the Proceedings, so you'll be able to peruse it at your convenience.

PHILOSOPHER: Professor Fregean, I was certainly glad I could be here today for your lecture, and I enjoyed it very much. I think that goes for all of us.

(FORMALIST, HISTORIAN, and LOGICIAN variously give their assent. FREGEAN nods and smiles.)

PHILOSOPHER: Since I really do wish to understand more about Frege, would each of you please tell me briefly what is most important to you about Frege's work? Historian?

HISTORIAN: Frege gave a symbolic language in which mathematical discourse could be effectively represented. This was a tremendous achievement.

PHILOSOPHER: Logician?

LOGICIAN: I agree with Historian, and I would add that in Frege's *Begriffsschrift* we can see the beginnings of the distinction between the formal language and the metalanguage—or in other words, that mathematical statements and proofs, formalized as in *Begriffsschrift*, can themselves become objects of mathematical investigation.

PHILOSOPHER: Professor Formalist?

FORMALIST: As Fregean and I like to say, quantification theory sprang full-blown from the head of Frege. And the people who lived in darkness saw a great light. ...if I might mix ancient metaphors this way.

PHILOSOPHER: Professor Fregean?

FREGEAN: Frege began a mode of thought that lies at the very heart and soul of analytical philosophy.

HISTORIAN: I think that it behooves us to recognize and even to celebrate great acts of discovery, as Leibniz said. André Weil gives the quote in an essay on the history of mathematics that I have with me... Yes, here it is: Leibniz said, "Its use is not just that History may give everyone his due and that others may look forward to similar praise, but also that the art of discovery be promoted and its method known through illustrious examples." (Weil, 1978, p. 227) Now, with this in mind, I would like to ask where Frege's work fits in the history of logic.

FORMALIST: That should be obvious. His work "fits", as you put it, at the very beginning of modern logic—thereby founding it.

FREGEAN: You might as well ask where Adam "fits" in the genealogy of the Old Testament!

LOGICIAN: What about Boole, De Morgan, Peirce, and Schröder? I have a copy of Hilbert and Ackermann with me, and they say... here it is: "The first clear idea of a mathematical logic was formulated by Leibniz. The first results were obtained by A. de Morgan... and G. Boole... The entire later development goes back to Boole." Then they go on to mention Peirce, Schröder, Frege, and Peano. (1950, pp. 1–2)

FORMALIST: As my very good friend, Willard Quine, has said, "Scholars engrossed in the logical tradition tend ... to place the beginning of modern logic in 1854 or earlier, rather than 1879, unaware of how much community there is in spirit and in substantive theory between Boole and De Morgan in the mid-nineteenth century and their mediæval and ancient predecessors." (1952, pp. VI–VII)

HISTORIAN: Did Quine actually mean to imply that *Hilbert* was too "engrossed in the logical tradition" to know what he was talking about?

FREGEAN: ... And as *my* very good friend, Michael Dummett, has said, "Boole cannot be called 'the father of modern logic.' ... anyone unacquainted with Boole's works will receive an unpleasant surprise when he discovers how ill-constructed his theory actually was and how confused his explanations of it." (1959, p. 205)

LOGICIAN: Anyone *acquainted* with Boole's works could easily tell that he was a first-rate mathematician! (Putnam, 1982, pp. 292–294)

FREGEAN: "... if (say) the development of logic had stopped short with Schröder, there would have been no clear sense in which contemporary logic would be superior to that of, for example, the Scholastics." (Dummett, 1959, p. 208)

LOGICIAN: But I really don't understand at all clearly what you mean by "modern logic". I think I have a pretty clear idea of what 'mathematical logic' means: on the one hand, 'logic'—whatever that is— treated by mathematical techniques; and on the other hand, the kind or kinds of logic used by mathematicians in their proofs. Historian and I have agreed that Frege's work was the first successful formal treatment of mathematical logic in the second sense.

HISTORIAN: In 'Logic as the Essence of Philosophy', Russell said that the "modern development of mathematical logic" began with Boole's 1854 book, although he acknowledged that Leibniz hoped for it and strove to develop it (1926, pp. 49–50). Later in Russell's essay, the phrase is shortened to "modern logic" (p. 68). Nevertheless, he also said that Boole had only the first sense—syllogistic logic treated with mathematical symbolism—and that Peano and Frege had made the first advance since the time of the Greeks toward the second sense—the logic applicable to mathematics (p. 50). Peirce had already treated the issue between these two senses of 'mathematical logic' very deeply in some of his unpublished manuscripts.

FORMALIST: Well, even the great Russell was capable of an occasional gaffe. Perhaps he needed a press secretary to exercise some control over the plebeian press.

FREGEAN: Yes, of course: What 'President' Russell really meant to say was, "Frege was the father of modern logic"; he only mentioned Boole and Peano by way of offering kindly consolations to them.

PHILOSOPHER: I thought that historians say the concept of 'modernity' arose in the seventeenth century, along with the nation-state and the scientific method. In fact, Jonathan Swift's 1704 essay, 'The Battle of the Books', was a mock-epic fable about the controversy over whether ancient or *modern* authors were better; and "the moderns" included such authors as Descartes and Hobbes (Swift, 1971, p. 388).

HISTORIAN: Yes, and that reminds me that Richard McKeon said in an essay with the same title as Swift's that Latin words like "ancient" and "modern" had already been used by fourteenth century logicians to distinguish the *antiqui*, who used only Aristotle's *Categories* and *On Interpretation*, from the *moderni*, who used the newly translated other four works of Aristotle on logic (1967, pp. 176–178, 185–186).

PHILOSOPHER: Then the word "modern" as meaning "newer and better" goes back even further than I thought—at least to the scholastic logicians! And they were "newer and better" because they were using all of Aristotle's *Organon*. The word "modern" seems to involve inevitable confusions, and even ironies, depending on who used it and when.

HISTORIAN: There is a profound difference between Boole's subject matter—which was traditional syllogistic logic—and his technical skill in drawing necessary conclusions or his creativity in using symbols to frame hypotheses and to make conjectures Boole's symbolic techniques would be unimaginable without the context of Descartes's and Leibniz's work, both of whom were "moderns" in Swift's sense. Even granting that monadic predicate calculus is a somewhat trivial—decidable, anyway—part of quantificational logic, Boole's work still led to remarkable advances in science: Surely Boolean algebra and switching theory as used in computer science and engineering are important developments. But when it comes to De Morgan, Peirce, and Schröder, there is *no way* that a calculus of relations can be construed as being a part of "scholastic logic"—and not even of '*moderni*' logic.

LOGICIAN: It seems to me that the statement, "Frege was the father of modern logic," is not so much a claim as a definition. If you define "modern logic" to be what Frege originated, then of course he was the founder of it.

PHILOSOPHER: If one says that Boole, De Morgan, Peirce, and Schröder were not "modern" because they didn't do what Frege did, doesn't that leave out the possibility that they might have been doing other things that were good in other ways?

FREGEAN: Yes, well, the philosophical layman can hardly comprehend or appreciate the vast importance of analytical philosophy as it arose out of Frege's work.

HISTORIAN: I have not studied analytical philosophy carefully, and it would therefore be arrogant of me to question your expertise by doubting whether analytical philosophy is vastly important or whether it arose out of Frege's work. But in my humble layman's opinion, much of scholasticism has far more in common with effete language games and endless disputes than it has with the serious work of Boole, Peirce, Schröder, Frege, Peano, and other real logicians. Perhaps we laymen will be able to console ourselves with circuit and switching theory arising out of Boole's work, with relational databases arising out of Peirce's and Schröder's work, and with Prolog and expert systems arising out of Peirce's, Herbrand's, and J. A. Robinson's work.

(FORMALIST and FREGEAN exchange knowing glances.)

FORMALIST: And we wish you the best in your consolations.

FREGEAN: Farewell.

(FORMALIST and FREGEAN leave; HISTORIAN, LOGICIAN, and PHILOSOPHER give each other blank stares.)

HISTORIAN: There go two promoters of logicism, intent as ever on bashing Boole and glorifying Frege.

PHILOSOPHER: Historian, you didn't want to be President of the American Philosophical Association anytime soon, did you?—say, within the next century or so?

(HISTORIAN laughs.)

LOGICIAN: You know, I hate to say this kind of thing, but I don't think the two of them were taking us seriously.

HISTORIAN: That's what we get for being revisionists. But perhaps there *is* something serious at stake here: Frege and Peano symbolized mathematical discourse. What can such work mean to non-mathematicians? And I don't mean that as being against you, Philosopher—nor as being against Formalist and Fregean.

PHILOSOPHER: No, Historian, I wouldn't have taken it that way. But I *can* see that there is something curious about their insistence on Frege's importance. Why should the person who succeeded in giving a language for mathematical proofs be so important to individuals who haven't cared to *do* many mathematical proofs? Without direct experience of mathematics, they must either be accepting his importance second hand or else seeing in his work something important that is not 'mathematical logic' in the sense of the logic used by mathematicians—or perhaps they even have some other kind of motive.

But Logician, you promised earlier that you would continue our discussion about understanding the completeness and incompleteness theorems...

LOGICIAN: Yes, well, it had occurred to me that everyone—everyone who lives in society, that is—has the experience of following rules and that most people who get an elementary education learn the algorithms for arithmetic with Hindu-Arabic numerals. So when a formal system of deduction rules is discussed, most people at least *feel* that they can understand what is being discussed.

PHILOSOPHER: That feeling of familiarity (that is, familiarity with following rules) would be Peirce's first grade of clearness for the idea. The abstract definition of a formal system would be the second grade of clearness. But for most people the experience of rules in society and—I'm sorry to say—in school mathematics seems to them quite arbitrary; and so the third grade of clearness, which would be the apprehension of the conceivable effects of the idea of a formal system, would elude most people.

LOGICIAN: That's very interesting... but conceivable effects could be either within the formal system itself or in relation to something external to it.

PHILOSOPHER: I'm certain that Peirce would have chosen the latter sense. Within the system itself 'truth' could only mean consistency, but Peirce emphasized that 'truth' must be the correspondence with an external reality—'external' in the sense that it does not depend on what any individual thinks.

HISTORIAN: And a model is an interpretation of the formal language in a set-theoretic reality which is external to the formal system.

LOGICIAN: Yes, exactly. To finish summarizing my thoughts about the question, it occurred to me that the proof of the incompleteness theorem is mostly syntactic or mostly about rules and that the popular accounts of the proof are entirely about rules. All the semantic content seems to have vanished from the following statement of undecidability: "There exists a proposition A such that neither A nor null-A can be derived in the formal system for arithmetic."

HISTORIAN: Yes, indeed. The word "incompleteness" still suggests that there is something *really* true that we can't prove, but the word "undecidable" makes no reference to an extra-syntactic reality.

PHILOSOPHER: I just thought of something: Is it reasonable to say that incompleteness and undecidability are not about whether any particular mathematician—or even all the mathematicians who have ever lived and ever will live—can't prove something, but rather about the fact that it absolutely cannot be done? That would be similar to Peirce's definitions of "external" and "reality", insofar as the fact wouldn't depend on what any individual thinks or is able to think.

LOGICIAN: Yes, the fact is mathematically *proved* about the particular formal system in the proof. And finally, the now popular Henkin proof of the *completeness* theorem is mostly about rules—at least in the sense that the model is constructed from the syntax of the formal language (Dreben and van Heijenoort, 1986, p. 57). This really got me curious, because it seemed to me that there is some kind of avoidance of semantics going on here. And yet, the completeness theorem is inescapably about the connection *between* syntax and semantics.

HISTORIAN: While it may take more effort or a different kind of effort to go through Gödel's original completeness proof, which uses the construction of partial models, the reward lies in a much greater understanding of the semantic or model-theoretic aspects of the theorem.

LOGICIAN: Now for me, the upshot of all this is that as a professional in logic I find the completeness theorem *easier* to understand than the incompleteness theorem. To me the completeness theorem is about something definite and concrete, whereas the incompleteness theorem is about something impossible.

Well, Historian, it's your turn to comment on my remarks, since you originally brought the subject up today. What do you think?

HISTORIAN: I take it that by "something impossible" you mean the impossibility of axiomatizing the theory of the standard model of arithmetic.

LOGICIAN: Of course.

HISTORIAN: Well, Logician, I think that you have the *right* philosophical outlook or epistemological attitude toward metamathematics and non-finitary reasoning, to coin a phrase from what Gödel said about Skolem. Nevertheless, there is a way to view Gödel's incompleteness proof as being about something definite and concrete: his title was, 'On formally undecidable propositions of *Principia Mathematica* and related systems I', and he actually constructed an undecidable proposition. Now this reminds me of Peirce's second grade of clearness to the extent that the undecidable proposition is formal and abstract; your concern with impossibility reminds me of Peirce's third grade of clearness—or perhaps *un*clearness in this case!—to the extent that you have gone on to consider the conceivable effects of the existence of undecidable propositions.

But we are in danger of being impolite to our good friend, Philosopher, again. Let's get back to your original question: Why does it seem that the incompleteness theorem is easier than the completeness theorem for non-professionals in logic to understand? You have adduced as circumstantial evidence the existence of popularizations of Gödel's incompleteness proof, and Philosopher exemplifies a non-professional reader of such popularizations, as he mentioned to us a couple of days ago. I can think of two reasons why we do not see any popularizations of the completeness theorem. First, it is possible at least to suggest what an undecidable proposition is by using natural language statements, but the concept of completeness really cannot be stated in ordinary language. (And I think this is related to what we were saying about Gödel's phrase, "the question at once arises.") What do you think, Philosopher?

PHILOSOPHER: Well, I did give a couple of paradoxical statements the other day, so that means they are memorable to me. But I also recall that the two books I've read—or tried to read—included symbols and examples of formal systems. I believe I mentioned that the books were *Gödel's Proof*, by Nagel and Newman (1958), and *Gödel, Escher, Bach*, by Hofstadter (1979).

HISTORIAN: I believe that purely verbal books about the Liar paradox are not useful except as amusements. But I also believe that the two books you chose are serious efforts, and I respect those authors for their good work. You chose wisely or fortunately.

(LOGICIAN grins.)

PHILOSOPHER: Actually, I asked Logician, and he recommended them to me.

(HISTORIAN laughs.)

HISTORIAN: Then it's gratifying to be in such good company—in more ways than one. Now my second reason why we do not see popularizations of the completeness theorem is less specific, but it occurred to me because of something you said, Philosopher. My experience in teaching mathematics convinces me that it is much harder to motivate students to take seriously a proof of what they already believe is obvious than it is to motivate proofs of what seems remarkable, surprising, interesting, etc. For example, to prove that between two points on a given line there exists another point on that line in a rigorous approach to the foundations of geometry is more difficult to motivate than is the Pythagorean Theorem; and to prove that $1 + 1 = 2$ in a rigorous (or even semi-rigorous) development of the foundations of arithmetic is more difficult to motivate than is the infinity of prime numbers. When you said, Philosopher, that you couldn't make sense out of the completeness theorem, you actually said that it was the idea of incompleteness that bothered you, didn't you?

PHILOSOPHER: Yes, and I think I see what you are getting at. Since it seems to me that completeness *ought* to be true, I find it difficult to experience a real doubt and therefore no need to try to 'fix' my belief—if you will excuse my making such a pun by way of allusion to Peirce.

HISTORIAN: In those terms, I guess the pedagogical problem I brought up as a reason we don't see popular accounts of completeness is that we cannot get students to experience a real doubt of something they already believe.

LOGICIAN: Historian, this second reason of yours is *also* related to what we were saying about Gödel's phrase, "the question at once arises."

HISTORIAN: You see, Philosopher, how Gödel's remark has caused Logician and me to experience real doubt—both about what Gödel actually meant and about what the completeness theorem means to us. And I've just now realized how we can tie this most recent line of inquiry back to what we experienced earlier today with Formalist and Fregean. I think that an individual whose intellect has been trained in the *a priori* method, as Peirce called it, is more concerned with consistency than with reality; Peirce said that the *a priori* method "makes of inquiry something similar to the development of taste" (1934, p. 241; 1986a, p. 253). For such an individual, the value of *Begriffsschrift* and *Principia Mathematica* would lie in providing a rigorous development of consistency—consistency being a deduction-theoretic criterion and correspondence with reality being, in a sense, a model-theoretic criterion.

PHILOSOPHER: Yes, Peirce said in many places that metaphysical philosophy had been characterized by endless disputes and that the *a priori* method was characterized by a desire for unique positions which could be easily defended in disputatious combat. But he also said that if metaphysics were pursued according to the method of science, it would be characterized by rapidly settled disputes, and it would progress through a sequence of problems solved and solutions accepted by a community of philosophical inquirers.

HISTORIAN: Well, for all I know, analytical philosophy *is* the kind of scientific philosophy that Peirce promoted; but if so, I would have expected Formalist and Fregean to try to help us to understand that and to try to acquaint us with such a community of philosophical inquirers. (In fact, the way they talked about Frege even reminds me of what Peirce called the method of authority.) Work like that of Peirce and Löwenheim involves truth as a kind of correspondence with the reality of different models; work like that of *Begriffsschrift* and *Principia* runs into problems insofar as the one enormous universe of all concepts or sets endangers the desired consistency. Furthermore, logicism makes mathematics subordinate to 'logic', which might appeal to logicians who wish to ignore mathematics proper. If logic is defined as "logistic", the way Alonzo Church did, one can understand that semantics and model theory would then actually belong to mathematics itself rather than to logic; and we can see that model-theoretic considerations are almost totally ignored in Church's *Introduction to Logic* (1956).

LOGICIAN: Oh, I see now... If one *defines* 'modern logic' to be the mathematical theory of formal deduction, then Frege really *is* the founder of it.

HISTORIAN: The incompleteness theorem can be treated with very little reference to model-theoretic notions, but the completeness theorem is *essentially* about the connection between syntax and semantics. The right philosophical outlook which I attributed to you, Logician, must necessarily be founded on a great deal of model-theoretic experience: one must have reached all three of Peirce's grades of clearness both with respect to syntax versus semantics and with respect to object language versus metalanguage. I think that we three have made some real progress in solving or at least understanding a problem in the history of mathematical logic. Don't you?

LOGICIAN: Well, however much progress we've made, it has certainly been interesting. And I think that our interaction with Formalist and Fregean was at least stimulating.

HISTORIAN: Oh, yes. I have reached a new clarity in my own ideas, and it probably would not have happened if you had not invited them to join our conversation.

PHILOSOPHER: Well, I may have merely increased my familiarity with the ideas of completeness and incompleteness, which is only the first grade of clearness, but you two have helped me to begin to understand the abstract definitions involved. And I've certainly gained a renewed interest in mathematics and the history of logic.

## 6.2 Why isn't Peirce's logical work more well-known?

In his essay, 'Set-Theoretic Semantics', Jean van Heijenoort came very close to defining the two lines of development in mathematical logic:

> When, after 1870, the development of modern logic began to take some impetus, there appeared two streams that, until 1920, hardly seem to mix their waters. One stream is that of Frege and Russell (to whom we should perhaps adjoin Peano), and the other is that of Peirce, Schröder and Löwenheim. We could say, as an initial approximation, that the first stream is syntactic, while the second is semantic. This is not quite correct, since Frege and Russell have their semantics, these semantics being different from each other and from set-theoretic semantics. Peirce, Schröder and Löwenheim consider domains of individuals, domains on which properties and relations are defined, and most of the time ignore formal proofs; their results, which today we would characterize as model-theoretic results, are obtained by semantic considerations. (van Heijenoort, 1977b, p. 183)

He implies—but does not state clearly—that the concept of derivability in a formal system unites the syntactic stream and that the concept of logical consequence unites the semantic stream. In Gregory Moore's review of *From Frege to Gödel: A Sourcebook in Mathematical Logic, 1879–1931* (which was edited by van Heijenoort) Moore was very critical of the fact that Boole, De Morgan, Peirce, and Schröder had been ignored in the sourcebook (Moore, 1977). Indeed, the very title of the sourcebook is an endorsement of the claim that Frege was the founder of "modern logic"; perhaps van Heijenoort was trying to make amends in the quotation above.

The leading idea of the deduction-theoretic line is derivability; the leading idea of the model-theoretic line is logical consequence. These ideas are so clear to us nowadays that it is difficult to read the works of Löwenheim and Skolem without anachronistically assuming that these ideas were clear to them and to their contemporaries. Especially, we need to realize that set-theoretic semantics rests on a knowledge of set theory that did not become commonplace until later. Carnap said in his autobiography that he had not even heard of Cantor and set theory, "which no professor had ever mentioned", until another student in Frege's summer 1913 'Begriffsschrift II' seminar told him (1963, p. 5). Skolem meant satisfiability and logical consequence when he used terms (such as "consistency") that are now used with respect to derivability. Tarski's early efforts to define logical consequence were actually concerned with derivability. Even Gödel did not adequately understand Skolem's viewpoint, because the words Skolem used misled Gödel to believe that Skolem had related syntax with semantics sufficiently to have actually stated (or nearly stated ) the completeness theorem in 1928 (Skolem, 1977c, p. 519; Wang, 1974, pp. 8–11).

So why is Peirce's work not better known? In the first place, with regard to his own time, there are difficulties arising from the mathematical content of Peirce's logic papers—difficulties which are essentially related to the Stage One analysis carried out in Chapters 2 and 3 of this book. While it is possible to use today's concepts to make Peirce's thought clear to present-day logicians, Peirce himself did not have the advantages of these concepts, especially set-theoretic ones. In his own time, the Stage One analysis we have given was really impossible. Even today, the necessary set-theoretic concepts—especially relations interpreted as subsets of Cartesian products—are only studied by students interested in pure mathematics or in mathematical logic, and then only at the advanced undergraduate or graduate level. And Peirce's use of example relations taken from human relations (as De Morgan did) introduced extra-mathematical ambiguities. Furthermore, Peirce did not make use of material implication as a connective in his quantificational logic. Peirce was quite adept at transforming ordinary language into his symbolism, especially by putting implications into disjunctive normal form, leaving out steps that must be filled in by the reader.

In the second place, there are also difficulties arising from Peirce's elliptical or even fragmentary style, which are related to the Stage Two analysis of his papers. But perhaps the preceding statement is not quite fair to Peirce. The appearance of being fragmentary or disjointed is not so much due to careless gaps. Peirce just tried to put too many ideas into his papers without getting these ideas into their most effective arrangement; these ideas often include philosophical or metaphysical notions that may seem to be extraneous to someone interested only in the mathematical aspect of logic. Certain symbols are deliberately used with more than one meaning within the same paper, and certain concepts are expressed in different symbols in different papers. The extenuating circumstances of Peirce's two professions—one as a scientist with the Coast and Geodetic

Survey and the other as a lecturer at Johns Hopkins—must have made it difficult for him to take the necessary time needed to revise his papers to be clearer and more rigorous. And he very likely felt some pressure to publish as much and as quickly as he possibly could, while he was still trying to secure a full-time position at Johns Hopkins.

In the third place, although the sociology or social psychology of the history of science is beyond the scope of this book, suffice it to say that Peirce was essentially blackballed from academic society (at least in America). He had many troubles over his apparent irreverence for religious authority and moral orthodoxy. He was finally removed from his post at Johns Hopkins because he had lived with his second wife, Juliette, before he was divorced from his first wife, Melusina Fay—a fact made known to the Trustees of Johns Hopkins by Simon Newcomb (Houser, 1986, pp. lxii–lxv). And as the new editor of the *American Journal of Mathematics*, Newcomb also suppressed the publication of the second installment of Peirce's great 1885 paper on mathematical logic, claiming that its subject was not mathematics (pp. xl–xli); some notes for the intended second part of the paper appear at (Peirce, 1933a, pp. 239–249). During the 1880s, Peirce had multiple difficulties at the Coast and Geodetic Survey; after Benjamin Peirce died in 1880, Peirce no longer enjoyed the influence that his father, who had been the Survey's first superintendent, could have provided. The Survey also came under intense criticism in the popular press as being a waste of money, and a number of Peirce's projects were curtailed or even cancelled. Peirce was more or less forced to resign his position in 1891. (Houser, 1986, pp. xx–xxiv, xxvii–xxxvi, xl–xli)

"Arisbe", the name Peirce gave to his 'retirement' home in Milford, Pennsylvania, was the name of a city near ancient Troy:

> **Aris′ba:** a city in the Troad near Abydos; according to Homer (*Iliad*, II, 836), it sent auxiliary forces to Troy. (Mandelbaum, 1981, p. 360)

This name has sometimes been explained as being an allusion to the wanderings of Aeneas, which was certainly a part of Peirce's meaning. Nevertheless, it is also an allusion to the following passage from Book 9 of the *Aeneid*:[5] Ascanius, the son of Aeneas, speaks to Nisus and Euryalus, two companions of Aeneas who have bravely proposed to go from where they are under siege to bring Aeneas to their aid. Ascanius says,

> "And I, whose only safety lies in my

---

[5]It is impossible to convey the bitterness, the irony, and even the hope which Peirce was suggesting through this complex allusion without providing an extensive discussion: of Peirce as Aeneas in exile from his destroyed home with the calamities of 1884–91 as the Trojan war; again of Peirce as Virgil hoping through his work to contribute to a great new age and despairing that the work was not yet complete; etc. The suggestion given above, which relies directly on the passage from the *Aeneid*, shall have to suffice.

dear father's coming back, beseech you both,
...
in you I now place all my hope and fortune;
recall my father, let me see Aeneas
again; with his return, all grief is gone.
And I shall give to you two silver bowls
skillfully wrought, embossed, both taken by
my father at the conquest of Arisba;
two tripods; two golden talents; and
an ancient goblet, gift of Sidon's Dido."

(Virgil, 1981, pp. 233–234)

The death of Peirce's father was not only personally devastating to him, but it was also politically and socially a very great loss; I believe it is this sentiment of missing his father that is uppermost in Peirce's allusion to this passage.

There has been a general impression that Peirce was a difficult man to get along with. He had many difficulties with publishers over proposed books, a number of which were actually in final manuscript form, including books on philosophy, on logic, and a series of school mathematics textbooks. In 1892, William James recommended Peirce for a position in philosophy at the newly formed University of Chicago; G. H. Palmer, a Harvard philosophy professor, wrote to President William Rainey Harper:

> I am astonished at James's recommendation of Peirce. Of course my impressions may be erroneous, and I have no personal acquaintance with Peirce. I know, too, very well his eminence as a logician. But from so many sources I have heard of his broken and dissolute character that I should advise you to make most careful inquiries before engaging him. I am sure it is suspicions of this sort which have prevented his appointment here, and I suppose the same causes procured his dismissal from the Johns Hopkins. (Houser, 1986, p. lxv)

Houser observes that,

> It is remarkable that James, certainly a man of judgment and discrimination, never gave up on Peirce but continued to recommend him both as teacher and scholar. Regrettably, others were blind to what James saw in Peirce. (Houser, 1986, p. lxv)

One can easily imagine how poorly Peirce would have gotten along with Palmer (and very likely did get along with professors like him) after comparing the following statement from Palmer's autobiography with Peirce's philosophical views as explained in Chapters 4 and 5 above:

127

> In Religion all Philosophy culminates, or rather, from it all
> Philosophy flows. To it and to its nearest of kin, Ethics, my life
> has been given. In the remainder of this paper I shall set forth
> as simply as possible the beliefs about the two to which the
> philosophic wanderings hitherto described have conducted me.
> And in discussing Religion I shall confine myself to Christianity
> as its universal type. For as we have it to-day it is all-inclusive
> and readily finds room within itself for the many precious
> half-truths of the other ethnic faiths. (Palmer, 1930, p. 73)

And indeed, even to this day [1992] Peirce is a non-person in Cambridge,
as his house is marked on the tourist maps, "Melusina Fay Peirce house".

Nevertheless, Joseph Jastrow, one of Peirce's students at Johns Hopkins,
gave a very positive account of Peirce in his memoir, 'Charles S. Peirce as
a Teacher':

> Mr. Peirce's personality was affected by a superficial reticence
> often associated with the scientific temperament. He readily
> gave the impression of being unsocial, possibly cold, more truly
> retiring. At bottom the trait was in the nature of a refined
> shyness, an embarrassment in the presence of the small talk
> and introductory salutations intruded by convention to start
> one's mind. His nature was generously hospitable; he was an
> intellectual host. In that respect he was eminently fitted to
> become the leader of a select band of disciples. Under more
> favorable circumstances, his academic usefulness might have
> been vastly extended. . .
>
> The young men in my group who were admitted to his circle
> found him a most agreeable companion. The terms of equality
> upon which he met us were not in the way of flattery, for they
> were too spontaneous and sincere. We were members of his
> "scientific" fraternity; greetings were brief, and we proceeded to
> the business that brought us together, in which he and we found
> more pleasure than in anything else. This type of cooperation
> and delegation of responsibility came as near to a pedagogical
> device as any method that he used. (Jastrow, 1916, p. 725)

Jastrow's description must be based on a reality; it is not the kind of thing
that could be made up merely for the sake of encomium or eulogy.

There is some circumstantial evidence of enmity between Peirce and
Russell. In *Semiotic and Significs: The Correspondence between Charles
S. Peirce and Lady Welby* edited by Charles Hardwick, there are several
letters from Peirce to Lady Welby in which he made harsh and negative
remarks about Russell and Whitehead (Peirce and Welby, 1977, pp. 9,
14, 28–30, 43). Lady Welby apparently showed or sent copies or versions
of at least some these remarks to Russell in 1904: she said that she had

done so (pp. 39, 41–42, 52), and Russell wrote to her about what Peirce had said (Hardwick, 1977, p. xxx; Peirce and Welby, 1977, p. 52)[6]. It is certainly possible that Peirce sent copies of some of his papers and manuscripts to Welby and that she shared them with Russell. As was mentioned in the dialogue above, there are ideas and phrases very similar to Peirce's in Russell's 1914 lectures, *Our Knowledge of the External World*, and especially in the second lecture, 'Logic as the Essence of Philosophy' (1926, pp. 42–69); but there is no reference to Peirce, of whom Russell was most certainly well aware. Russell's lectures were delivered in Boston during March and April of 1914 as the Lowell Lectures, a venue at which Peirce had more than once given lectures of his own, including the 1903 lectures on pragmatism. During the time Russell was giving these lectures, Peirce died in poverty and obscurity. Russell wrote in his forward to *An Introduction to the Philosophy of Charles S. Peirce*, a synthesis of Peirce's philosophy by James Feibleman,

> Peirce was a man of tremendous energy, producing a multitude of ideas, good, bad, and indifferent. He reminds one of a volcano spouting vast masses of rock, of which some, on examination, turn out to be nuggets of pure gold. (Russell, 1946, p. xvi)

Indeed! Perhaps Russell's eminence and Whitehead's position at Harvard have had something to do with the ill fortune of Peirce's reputation.

Finally, with respect to 'modern' times, the dialogue in Section 6.1 above makes some suggestions as to why Peirce is not better recognized by present-day writers on logic. Some authors have been particularly concerned to diminish the importance of Boole and De Morgan, for example (Dummett, 1959) and (Quine, 1951; 1952; 1985). Quine's reviews of volumes 3 and 4 of *Collected Papers* are particularly interesting in this regard (1934a; 1934b). There seems to have been a controversy in recent years over the importance of Boole, Peirce, Schröder, Löwenheim, and Skolem, of which there are only a few indications in the literature. The philosopher and logician Hilary Putnam was stimulated by Quine's attitude toward Boole to write a rejoinder, 'Peirce the Logician', defending Boole and Peirce (1982, pp. 292–295).[7] Most of the following items use an almost identical commonplace of "Boole-Peirce-Schröder-Löwenheim-Skolem", sometimes shortened but usually including the hyphens: (Dreben and van Heijenoort, 1986, p. 44), (Fisch, 1986b, pp. 437–438), (Goldfarb, 1971, p. 2; 1979, pp. 354–356), (Moore, 1977, p. 469; 1988, pp. 96–100, 102–104), (Putnam, 1982), (van Heijenoort, 1977b, p. 183), (Wang, 1987, p. 266). Moore, Fisch, and Putnam hold out for the importance of Boole's and Peirce's

---

[6]The account given above is not quite the same as Hardwick's; there appears to be either a misprint or a mistake at (Hardwick, 1977, p. xi) and at (Peirce and Welby, 1977, p. 39).

[7]It appears that Quine was chastened by this 1982 criticism, so that his use of the "sprang from the head" image was more carefully qualified in (Quine, 1985).

work; the others are somewhat ambivalent or perhaps even confused about the significance of the entire line. But De Morgan's role is almost never mentioned; and although some of them mention semantics, they do not explain that the elements of model theory were developed in the De Morgan-"Boole-Peirce-Schröder" line or stream.

Although Peirce is recognized in diverse places for diverse and apparently fragmentary contributions, the idea of a model-theoretic line of development has not been used as the single unifying concept to acknowledge what he achieved and to place his work correctly in the history of logic.

# Chapter 7

# Further Developments of Peirce's Ideas

Toward the end of Peirce's 1878 paper, 'How to Make Our Ideas Clear', he had this to say about the importance and influence of ideas:

> We have, hitherto, not crossed the threshold of scientific logic. It is certainly important to know how to make our ideas clear, but they may be ever so clear without being true. How to make them so, we have next to study. How to give birth to those vital and procreative ideas which multiply into a thousand forms and diffuse themselves everywhere, advancing civilization and making the dignity of man, is an art not yet reduced to rules, but of the secret of which the history of science affords some hints. (Peirce, 1934, p. 271; 1986a, pp. 275–276)

In this concluding chapter, we sketch some developments of Peirce's logical ideas in present-day computer science and discuss some of Peirce's ideas about logic machines, reasoning, and education.

## 7.1 Today's first-order logic

Where did today's first-order logic come from? This question was implied but not answered in the dialogue of Chapter 6. First-order logic can be traced back to Hilbert and Ackermann's book (1950); they streamlined the syntax of Russell and Whitehead's *Principia Mathematica* and concentrated on the first-order part of logic. There appear to be two sources for the discussion of first-order logic in Hilbert and Ackermann: In the first place, their concern with the relation between universal validity and provability, as a result of work in the 1920s on the decision problem, originated from the model-theoretic approach in Löwenheim's paper of 1915 (Dreben and van Heijenoort, 1986, p. 47). In the second place, Hilbert gave a course of lectures on logic in the winter of 1917–18 (Moore, 1988, pp. 96, 113–116); in these notes the logic is almost entirely formal and deductive. Set-theoretic semantics is essentially not mentioned in Hilbert's lectures; instead, he used an informal semantics based on mathematical statements that have truth values with respect to mathematical experience. If we assume that we already know what truth is in mathematics, then we can translate a formal statement into an informal statement about mathematics and consider the truth value of that informal statement.

# 7.2 Practical consequences in computing

In this section we consider: the development of Peirce's relational algebra of 1870 into the current use of relational data bases in information technology; the line of development from his mathematical logic of 1885 to the computer language Prolog and expert system applications; and two developments from other work of his which was not investigated in this book. In relational data bases and expert systems, the conceivable effects of Peirce's ideas have been borne out in important practical consequences.

## 7.2.1 Relational databases

Today's concept of a relational data base was developed ultimately from Peirce's 1870 relational algebra—but probably via Russell and Whitehead— by E. F. Codd exactly one century later (Codd, 1970). One of the most important features of the relational model for a data base is the definition of new relations; once defined, the new relations are just as usable as the original ones. In other words, we have the second-intentional logic: The relations that were thoughts have become things; extensionally the new relations are ordered sets of n-tuples. There are many relational data base software packages now available, and Structured Query Language is the high level language for controlling these relational data bases.

Codd's initial insight was to use the first-order predicate calculus as a query language for data base applications (1970). Thus, there was already an influence of Hilbert and Ackermann on Codd's ideas. This insight led directly to his design of an appropriate data base model for such first-order queries—namely, the relational model. Codd tells an anecdote about the lack of knowledge of mathematical logic amongst people working in the field of data bases:

> I began working in data bases in August of 1968. Just before then I was in Poughkeepsie, N.Y., where I attended a talk given by a speaker from a company beginning to market a data base system. I meant to ask him two questions. The first concerned the existential quantifier of logic. I asked him to

what degree the product supported it. I was going to ask about the universal quantifier, but by the way he answered the first question I knew he knew nothing about predicate logic. He said, "I often get questions of a philosophical nature, but this is the first time I've had a question pertaining to existential philosophy." I thought, "what is this data base field doing if the product designers don't know anything about predicate logic?" I feel predicate logic is an essential tool. (Codd, 1988, p. 89; 1990, pp. 35–36)

Codd's join operation is of the same nature as Peirce's relational product, although in the join the intermediate place is retained in the resulting relation (1970; 1987)—for example, the relational product of two binary relations is also a binary relation, whereas the join of two binary relations is a ternary relation. As Codd himself stressed, the key to working with the relational model is to treat relations as objects:

With data bases, a lot of concern is with the relationship between things. Why not treat the aggregate unit as a relation and then say that some of these relations represent merely the relationship between an object and its properties? Then, when you speak of relationships between objects, you can have properties of these relationships because the relationship itself is an object. (Codd, 1988, p. 90)

This is obviously in the tradition of Peirce and Schröder, insofar as set-theoretic semantics is necessary to such a conception. Even though Codd was probably not directly aware of Peirce's and Schröder's work, the relational formulas in Russell and Whitehead were almost certainly adopted from Schröder's work, as was mentioned in Chapter 6.

There is a remarkable irony here: IBM sent Codd to the University of Michigan, where his computer science thesis committee included Arthur Burks (Codd, 1965, p. ii). Burks was a pioneer in computer science, but he was also the Peirce scholar who edited volumes 7 and 8 of the *Collected Papers* (Peirce, 1958a; 1958b). However, according to Burks[1], Codd did not study Peirce or Peirce's relational algebra at Michigan (Burks, 1989, personal communication). Codd's thesis is not about relations or relational algebras (Codd, 1965).[2]

---

[1] This material is from a conversation between Burks and the author in September 1989.

[2] It has been an obvious inference to make, that there was a connection between Codd and Peirce mediated by Burks—made by others besides me. Hence I was surprised by what Burks had to say. about it.

### 7.2.2 Prolog and expert systems

Jacques Herbrand, in developing his rules and the concept of the "Herbrand universe", was building on the work of Löwenheim and Skolem (as well as on work in the deductive-theoretic tradition) and thus ultimately on the work of Peirce (Herbrand, 1971). In about 1960, a number of logicians, including Martin Davis, Hilary Putnam, and Hao Wang, worked at applying Herbrand's ideas to theorem-proving on computers (Robinson, 1963, p. 174). Building on their work, J. A. Robinson invented the resolution method in his 1965 paper, 'A Machine-Oriented Logic Based on the Resolution Principle'. Prolog originated in 1972 from a combination of work at the University of Edinburgh, especially by R.A. Kowalski, on computer theorem-proving and work at the University of Marseilles, especially by Colmerauer and Cohen, on compiler theory and design. See the three historical accounts (J. Cohen, 1988; Kowalski, 1988; Robinson, 1992). The line of development from Peirce to Prolog is thus objectively established through the use of ideas by the logicians in the line: Peirce–Schröder–Löwenheim-Skolem-Herbrand-Robinson-Prolog.

### 7.2.3 Other developments

After 1885, Peirce turned his attention to a new graphical method of developing logic. The result was his theory of existential graphs, which he worked out in many published papers and unpublished manuscripts. Peirce's graphs have been studied almost exclusively by Peirce scholars (Roberts, 1973; Zeman, 1964). Recently, John Sowa has written a book that utilizes Peirce's existential graphs in computer processing of natural language (1984). Peirce himself said in a 1908 article,

> In reference to those [existential] graphs, it is to be borne in mind that they have not been contrived with a view to being used as a calculus, but on the contrary for a purpose opposed to that. Nevertheless, if anyone cares to amuse himself by drawing inferences by machinery, the graphs can be put to this work, and will perform it with a facility about equal to that of my universal algebra of logic and as much beyond that of my algebra of dyadic relatives, of which the lamented Schröder was so much enamoured. (Peirce, 1933b, pp. 513–514)

Perhaps Sowa's work is the beginning of an important practical application of Peirce's existential graphs.

Recently there has been some work in artificial intelligence on implementing Peirce's concept of "abduction" as a form of inference. Abduction (also called "retroduction" and "hypothesis" by him) is a third form of inference separate from deduction and induction. 'Deduction, Induction, and Hypothesis', Peirce's last paper in his *Popular Science Monthly* series of 1877–78, contains a discussion of the three forms (1932, pp. 372–388; 1986a, pp. 323–338). A Spring Symposium on "Automated Abduction" was held in 1990 by the American Association for Artificial Intelligence; see (O'Rorke, 1990) for a report and a list of references. Several books have appeared, including *Abductive Inference Models for Diagnostic Problem Solving* (Peng and Reggia, 1990), which was reviewed by Thagard (1991). An article has even appeared in a popular magazine (Drake and Hess, 1990). Peirce would probably not have thought of his concept of abduction as something that could be automated, and it is not clear to what extent Peirce's word is being used correctly. But at least the germ of the idea is attributed to him in this artificial intelligence work.

# 7.3 Peirce on logic machines and creativity

## 7.3.1 Logic machines

In his 1887 article, 'Logical Machines', Peirce discussed the logical machines of Marquand and Jevons. Peirce's student Marquand had written a paper on a three term logic machine for the 1883 volume, *Studies in Logic*. Peirce especially discussed Marquand's four term machine, mentioned in an appended note to Marquand's paper (1883, p. 16). In his 1887 article, Peirce explained his own view of reasoning machines:

> The secret of all reasoning machines is after all very simple. It is that whatever relation among the objects reasoned about is destined to be the hinge of a ratiocination, that same general relation must be capable of being introduced between certain parts of the machine. ... This is the same principle which lies at the foundation of every logical algebra; only in the algebra, instead of depending directly on the laws of nature, we establish conventional rules for the relations used. When we perform a reasoning in our unaided minds we do substantially the same thing, that is to say, we construct an image in our fancy under certain general conditions, and observe the result. In this point of view, too, every machine is a reasoning machine, in so much as there are certain relations between its parts, which relations involve other relations that were not expressly intended. A piece of apparatus for performing a physical or chemical experiment is also a reasoning machine, with this difference, that it does not depend on the laws of the human mind, but on the objective reason embodied in the laws of nature. Accordingly, it is no figure of speech to say that the alembics and cucurbits of the chemist are instruments of thought, or logical machines. (Peirce, 1887, p. 168)

And in the same paper, Peirce conceived of a machine that would make calculations in the logic of relations:

> ...I do not think there would be any great difficulty in constructing a machine which should work the logic of relations with a large number of terms. But owing to the great variety of ways in which the same premises can be combined to produce different conclusions in that branch of logic, the machine, in its first state of development, would be no more mechanical than a hand-loom for weaving in many colors with many shuttles. The study of how to pass from such a machine as that to one corresponding to a Jacquard loom, would be very likely to do very much for the improvement of logic. (Peirce, 1887, p. 170)

Note well that the Jacquard loom was controlled by a chain of punched cards, tied together with string, which controlled the loom to make a certain pattern. And that gave Stanley Hollerith the inspiration to create a company to make the tabulating machines using punch cards that later became part of the IBM Corporation.

In 1886 Peirce had written about logic machines in a letter to Marquand, as mentioned by Martin Gardner in his book, *Logic Machines and Diagrams*:

> ...Not until the early 1970s did a letter come to light that Peirce had sent Marquand in 1886. After expressing belief that Marquand's machine could be improved to handle "very difficult problems," Peirce added: "I think electricity would be the best thing to rely on." Peirce then actually sketched circuits for both conjunction and disjunction, the first known effort to apply Boolean algebra to the design of switching circuits! (Gardner, 1982, p. 116)[3]

In the 1887 paper on logical machines, Peirce used conjunctive normal form and disjunctive normal form (pp. 166n1, 168–170). Toward the end of his 1885 paper, Peirce made a start at developing some Herbrand-style rules of deduction, but this was too difficult a task for him to carry out.

These ideas of Peirce's had no discernible effect on the development of computer science. As was the case with many of the remarkable ideas of Leonardo da Vinci, Peirce's ideas had to be rediscovered many years later. Fifty years after Peirce drew the circuits for 'and' and 'or' and conjectured their use in a logic machine, Claude Shannon established a complete implementation of Boolean algebra in circuits. For a few decades in the mid-twentieth century, punch cards, as Peirce conjectured, controlled the majority of large computers.

---

[3]See (Burks, 1975, pp. 303–304) for Peirce's letter.

## 7.3.2  Theorematic and corollarial reasoning

In Chapter 4 we discussed Peirce's view that the mathematician performs experiments on schemata; these experiments are real—the rapid settling of disputes in mathematics is evidence that these schemata employed in theorematic reasoning do not depend upon what any particular mathematician wishes or thinks. A logic capable of describing such schemata would itself need to be diagrammatical and iconic.

Boole's 1847 work, *A Mathematical Analysis of Logic*, was, in part, a contribution towards settling a dispute between De Morgan and Sir William Hamilton[4] and, in part, a response to Hamilton's ludicrous claim that the study of mathematics is harmful to the intellect. Peirce made a distinction between theorematic (or theoric) and corollarial reasoning in mathematics, which was at least partly motivated by his desire finally to settle the dispute between Boole and Hamilton. The controversy between De Morgan and Hamilton surrounded the "quantified predicate."[5] Hamilton introduced into syllogistic propositions the words "all" and "some" before the predicate as well as before the subject: for example, "All A is All B", "All A is Some B", and moreover with "Not" present or absent before each of the two quantifiers, etc.

Hence, from 'All A is All B'
through 'Some A is Some B',
through 'Not All A is All B',
finally to 'Not Some A is Not Some B'.
There are sixteen of these forms, several of which are utterly nonsensical. See (Heath, 1966) for a full account.

But it is Hamilton's attack on mathematics as a part of liberal education which concerns us here because Boole's involvement provided a context and a motivation for Peirce's distinction. Boole's remark about associating logic with mathematics instead of with metaphysics, quoted at the start of Section 4.1, was directed at Hamilton. The title of Boole's 1854 book, *An Investigation of the Laws of Thought*, has been a source of puzzlement to many. Russell contributed to this confusion in his essay, 'Mathematics and the Metaphysicians':

> Pure mathematics was discovered by Boole, in a work which he called the *Laws of Thought* (1854). This work abounds in asseverations that it is not mathematical, the fact being that Boole was too modest to suppose his book the first ever written on mathematics. He was also mistaken in supposing that he was dealing with the laws of thought: the question how people actually think was quite irrelevant to him, and if his book had

---

[4]A Scottish philosophy professor (1788–1856) at the University of Edinburgh who is not to be confused with Sir William Rowan Hamilton (1805–65), the Irish mathematician.

[5]This may have suggested the word "quantifier" to Peirce

really contained the laws of thought, it was curious that no one should ever have thought in such a way before. His book was in fact concerned with formal logic, and this is the same thing as mathematics. (Russell, 1971, p. 59)

As a matter of fact, Boole was very concerned with "how people actually think"—see (Laita, 1980). But the term "laws of thought" was simply Hamilton's definition of logic: "Logic—the science of the formal laws of thought" (Hamilton, 1853, p. 121); "the *laws of thought*, and not the *laws of reasoning*, constitute the adequate object of the science [of logic]" (p. 136).

Hamilton's attack on mathematics was actually disguised as a review of a pamphlet by William Whewell on the place of mathematics in liberal education. Here follow some representative passages from Hamilton:

> The study of Language, if conducted upon rational principles, is one of the best exercises of an applied Logic. This study I can not say that any of our universities encourage. To master, for example, the Minerva of Sanctius with its commentators is, I conceive, a far more profitable exercise of mind than to conquer the Principia of Newton.—But I anticipate. (Hamilton, 1853, p. 262n1)

> If we consult reason, experience, and the common testimony of ancient and modern times, none of our intellectual studies tend to cultivate *a smaller number of the faculties, in a more partial or feeble manner, than mathematics.* This is acknowledged by every writer on education of the least pretension to judgment and experience... (p. 268)

> 'No one, almost,' says Cicero, 'seems to have intently applied himself to this science [of mathematics], who did not attain in it any proficiency he pleased;'... (p. 280)

There is no explicit attack on De Morgan in Hamilton's review, but the review is the source for most of the references to Hamilton in Boole's *A Mathematical Analysis of Logic* (1847). Hamilton explained the difference between mathematics and philosophy thusly:

> In Mathematics we always *depart from the definition*; in Philosophy, *with the definition we usually end.*—Mathematics *know nothing of causes*; the *research of causes* is Philosophy; the former display only the that ($\tau o\ o\tau\iota$) ; the latter mainly investigates the why ($\tau o\ \delta\iota o\tau\iota$). (Hamilton, 1853, p. 273)

In his 1847 book, Boole referred to the preceding passage from Hamilton's review:

But the question before us has been argued upon higher grounds. Regarding Logic as a branch of Philosophy, and defining Philosophy as the "science of a real existence," and "the research of causes," and assigning as its main business the investigation of the "why, (το διοτι)," while Mathematics display only the "that, (το οτι)," Sir W. Hamilton has contended, not simply, that the superiority rests with the study of Logic, but that the study of Mathematics is at once dangerous and useless. (Boole, 1847, p. 11)

Peirce alluded to this dispute in the manuscript containing his definitions of mathematics and logic, part of which was discussed in Section 4.2.2 above:

But mathematics, as a serious science, has, over and above its essential character of being hypothetical, an accidental characteristic peculiarity—a *proprium*, as the Aristotelians used to say—which is of the greatest logical interest. Namely, while all the "philosophers" follow Aristotle in holding no demonstration to be thoroughly satisfactory except what they call a "direct" demonstration, or a "demonstration why"—by which they mean a demonstration which employs only general concepts and concludes nothing but what would be an item of a definition if all its terms were themselves distinctly defined—the mathematicians, on the contrary, entertain a contempt for that style of reasoning, and glory in what the philosophers stigmatize as "mere" indirect demonstrations, or "demonstrations that." (Peirce, 1933b, p. 193)

It was thus natural for Hamilton to glorify the "why" and to diminish the "that". Peirce then gave his own analysis of the difference between these two kinds of reasoning:

Those propositions which can be deduced from others by reasoning of the kind that the philosophers extol are set down by mathematicians as "corollaries." That is to say, they are like those geometrical truths which Euclid did not deem worthy of particular mention, and which his editors inserted with a garland, or corolla, against each in the margin, implying perhaps that it was to them that such honor as might attach to these insignificant remarks was due. In the theorems, or at least in all the major theorems, a different kind of reasoning is demanded. Here, it will not do to confine oneself to general terms. It is necessary to set down, or to imagine, some individual and definite schema, or diagram—in geometry, a figure composed of lines with letters attached; in algebra an array of letters of

139

which some are repeated. This schema is constructed so as to conform to a hypothesis set forth in general terms in the thesis of the theorem. Pains are taken so to construct it that there would be something closely similar in every possible state of things to which the hypothetical description in the thesis would be applicable, and furthermore to construct it so that it shall have no other characters which could influence the reasoning. (Peirce, 1933b, pp. 193–194)

In other words, theorematic reasoning requires that the mathematician imagine an individual schema or diagram or *model*. Then the logical consequences of the premises, as they are embodied in the schema or model, are investigated in a kind of experimentation:

How it can be that, although the reasoning is based upon the study of an individual schema, it is nevertheless necessary, that is, applicable, to all possible cases, is one of the questions we shall have to consider. Just now, I wish to point out that after the schema has been constructed according to the precept virtually contained in the thesis, the assertion of the theorem is not evidently true, even for the individual schema; nor will any amount of hard thinking of the philosophers' corollarial kind ever render it evident. Thinking in general terms is not enough. It is necessary that something should be DONE. In geometry, subsidiary lines are drawn. In algebra, permissible transformations are made. Thereupon, the faculty of observation is called into play. Some relation between the parts of the schema is remarked. But would this relation subsist in every possible case? Mere corollarial reasoning will sometimes assure us of this. But, generally speaking, it may be necessary to draw distinct schemata to represent alternative possibilities. Theorematic reasoning invariably depends upon experimentation with individual schemata. We shall find that, in the last analysis, the same thing is true of the corollarial reasoning, too; even the Aristotelian "demonstration why." Only in this case, the very words serve as schemata. Accordingly, we may say that corollarial, or "philosophical" reasoning is reasoning with words; while theorematic, or mathematical reasoning proper, is reasoning with specially constructed schemata. (Peirce, 1933b, pp. 194)

In other words, corollarial reasoning is purely deductive and mechanical, whereas theorematic reasoning is constructive and creative. Peirce was concerned to explain through this distinction how mathematics can appear to be deductive and yet present a continual unfolding of just as many startling discoveries as any other science. In an article of 1908, Peirce discussed these ideas again (1933b, pp. 505–513). He sharpened his terminology somewhat, replacing "theorematic" with "theoric":

> I shall term the step of so introducing into a demonstration a new idea not explicitly or directly contained in the premisses of the reasoning or in the condition of the proposition which gets proved by the aid of this introduction, a *theoric* step. ...Now to propositions which can only be proved by the aid of theoric steps...I propose to restrict the application of the hitherto vague word *"theorem,"* calling all others, which are deducible from their premisses by the general principles of logic, by the name of *corollaries*. (Peirce, 1933b, p. 509)

> I wish a historical study were made of all the remarkable theoric steps and noticeable classes of theoric steps. I do not mean a mere narrative, but a critical examination of just what and of what mode the logical efficacy of the different steps has been. Then, upon this work as a foundation, should be erected a logical classification of theoric steps, and this should be crowned with a new methodeutic of necessary reasoning. (p. 510)

And this might well be as close as Peirce ever got to describing model theory.

## 7.4 Peirce and education

There are four phases to be discussed here: Peirce as a student, both in school and college and in being tutored by his father; Peirce's vision of a mathematical logic; Peirce as a professor at Johns Hopkins, teaching logic and directing research; and Peirce as the author of school mathematics textbooks.

### 7.4.1 Peirce as a student

Peirce's father, Benjamin, was the greatest mathematician in the United States of his time, and he gave tremendous attention to making Charles into a prodigy. He invented complicated demonstrations with ordinary objects, such as playing cards, and would sometimes stay up all night taking Charles through these exercises. By the time Peirce entered Harvard College, he was working at graduate level algebra. In fact, he had learned a kind of 'new math' from his father—both in the sense that it included different number systems and abstract algebraic structures and in the sense that he learned mathematics the way it is understood by a research mathematician.

When Peirce was a boy, logic was in a moribund state. Peirce read Whately's *Elements of Logic* in the fall of 1851, when he was only twelve years old, and thereupon decided to devote his life to logic (Fisch, 1982, p. xviii; Peirce and Welby, 1977, p. 85). Peirce's account of his father's definition of mathematics was discussed in Section 4.2.2 above. Benjamin had comments about philosophical reasoning as well:

> Before I came to man's estate, being greatly impressed with Kant's Critic of the Pure Reason, my father, who was an eminent mathematician, pointed out to me lacunæ in Kant's reasoning which I should probably not otherwise have discovered. (Peirce, 1931, p. 299)

Eventually Charles collaborated with his father and edited some of his father's work. It is remarkable that in trying to reconcile philosophical schooling at Harvard with instruction in mathematics from his father, Peirce had to recapitulate the intellectual revolution of the new mathematical logic in his own mental development.

Peirce's philosophy teacher at Harvard, Francis Bowen, was a student of the philosopher Hamilton (discussed in Section 7.3.2 above) and a promoter of Hamilton's ideas (Bowen, 1873). Whately's book on traditional logic was still the text during Peirce's junior year at Harvard (Fisch, 1982, p. xix). Peirce never completely renounced this non-mathematical traditional logic that he encountered in the works of Whately and Hamilton; it always remained a part of his outlook. Nevertheless, by the time Peirce gave his first Lowell Lecture on the Logic of Science in 1866, he was able to say of Whately:

> But, indeed, few persons who have not had some special interest in the subject [of logic] are at all aware of the immense progress which has been made in the science during the present century. School-books are usually antiquated; and those who know nothing later than Whately, might as well judge Chemistry by the [books] which they find in their grandfathers' libraries as judge modern logic by that. (Peirce, 1982, pp. 359–360)

The situation of undergraduate education at Harvard was described by G. H. Palmer in his autobiography (1930). Palmer was the Harvard philosophy professor who recommended in 1892 that Peirce be denied a position at the University of Chicago, as was mentioned in Section 6.2 above. Palmer entered Harvard College in 1860 (1930, p. 11), the year after Peirce graduated, and he described it thus:

> Harvard education reached its lowest point during my college course. When I entered, it was a small local institution with nine hundred and ninety-six students in all its departments and thirty teachers in the College Faculty. ... Nearly all its studies were prescribed, and these were chiefly Greek, Latin, and Mathematics. There was one course in Modern History, one

in Philosophy, a half course in Economics. There was no English Literature. . . A feeble course or two in Modern Languages was allowed to those who wished it. There were two or three courses in Natural Science, taught without laboratory work. All courses were taught from textbooks and by recitations. . . . Professor Cooke, it is true, lectured to the Sophomores an hour each week on Chemistry. But though we were all required to attend, there was no examination. All teaching was of a low order. (Palmer, 1930, pp. 12–13)

Note that "Natural Science" was actually "taught without laboratory work"—and chemistry even without examinations. Palmer continues,

> . . . What I most wanted from Harvard was systematic training in Philosophy. But Professor Bowen offered only a single course and that more elementary than any of the more than thirty now on the Harvard list [that is, in 1930]. A slender acquaintance with the Scotch School—Reid, Stewart, Hamilton—was something. (Palmer, 1930, p. 14)

> . . . In most colleges the little Philosophy attempted was usually taught by the President, a minister. If an independent teacher was employed, he also was a minister. Under Puritanism Theology and Philosophy were pretty closely identified. Before the days of Johns Hopkins, too, the best opportunity for continuous study of Philosophy was in a Divinity School. (p. 20)

The traditional book of Whately was eventually replaced by Jevons's *Elementary Lessons in Logic*. Peirce found Jevons's book to be worthless (Peirce, 1986a, p. 4); even Palmer did not appreciate it, although this was possibly because it contained less "Philosophy" than Whately's:

> At the beginning of my teaching in Philosophy, I was merely the Assistant of Professor Bowen. He directed all my work, even what books my classes should use. The Heads of Departments in those days took their positions seriously. Logic was required of all Juniors and I was set to teach it in Jevons' Elementary Lessons. The class was divided into six sections, each of which I was to meet twice a week. Anything less nutritive can hardly be imagined. Jevons has carefully eliminated all Philosophy from his clever and shallow little book, so that my twelve hours a week of work would seem to have been completely unprofitable to me and to my class. (Palmer, 1930, p. 41)

## 7.4.2 The vision of a mathematical logic

We have already referred to Leibniz's vision of a mathematical logic at the beginning of Chapter 4. A similar vision, however, seems to have appeared

both to Boole and to Peirce. Desmond MacHale, Boole's biographer, writes
that Boole had such a vision when he was only eighteen years old:

> It was during his stay at Doncaster, early in 1833, that Boole
> first contemplated the ideas which were to grow into his major
> contribution to mathematics—the expression of logical rela-
> tions in symbolic or algebraic form. He relates that the thought
> flashed upon him suddenly one afternoon as he was walking
> across a field, but he laid it aside for many years, being inter-
> ested in other pursuits. The thought however smouldered in his
> subconscious and became an integral part of his main ambition
> in life—to explain the logic of human thought and to delve
> analytically into the spiritual aspects of man's nature. It was
> not until 1847 however that Boole—provoked and inspired by a
> controversy between the great logicians Sir William Hamilton
> and Augustus de Morgan—had sorted out his ideas and suf-
> ficiently developed his theory of symbolic logic to publish his
> views. The circumstances of his first thoughts on the possible
> connections between algebra and logic suggest that he had a
> vivid experience, somewhat like that of Saul on the road to
> Damascus or Descartes in the celebrated incident of the stove
> at Ulm. Boole referred to the incident many times in later
> life and seems to have regarded himself as cast in an almost
> messianic role. (MacHale, 1985, p. 19)

An account of Mary Everest Boole's description of her husband's vision
and his attitude to logic is given by Laita (1980, pp. 37 41). Thus, to
follow MacHale's account, Boole seems to have envisioned such effects as
settling disputes in metaphysics or even in theology through the expression
of the laws of thought in a symbolic logic. It does not seem likely that
Boole knew of Leibniz's attempts at algebraic logic until late in life (Laita,
1980, p. 58).

As we saw in Section 2.2 above, Peirce discussed Boole's work for the
first time in his 1865 Lectures on British Logicians and first wrote about
Boole's algebra of classes in a paper of 1867. But it does not seem that
Boole's work had affected Peirce very deeply by then, since most of Peirce's
other lectures and papers of that time dealt with the traditional logic. The
turning point appears to have occurred upon his study of De Morgan's
1860 paper on the logic of relations. In the following passage (which was
already quoted in Section 2.2 above) Peirce recalled, sometime about 1905
(1931, p. 299n), the effect that De Morgan's paper had on him:

> ...I at once fell to upon it; and before many weeks had come
> to see in it, as De Morgan had already seen, a brilliant and
> astonishing illumination of every corner and every vista of logic.
> ...his was the work of an exploring expedition, which every day

comes upon new forms for the study of which leisure is, at the moment, lacking, because additional novelties are coming in and requiring note. He stood indeed like Aladdin (or whoever it was) gazing upon the overwhelming riches of Ali Baba's cave, scarce capable of making a rough inventory of them. (Peirce, 1931, p. 301)

Peirce's own papers repeatedly take up new forms and additional novelties, also suggesting to us the metaphor of a report from an expedition into a new world of logic based on mathematics. Perhaps this vivid metaphor is an indication that Peirce had experienced his own vision of a mathematical logic, brought about by his work on the ideas in the papers of De Morgan and Boole (and possibly even Leibniz). In the article of 1908, quoted at the end of Section 7.3.2 above, Peirce implied that he had actually experienced the vision himself:

My future years—whatever can have become of them, they do not seem so many now as they used, when, at De Morgan's *Open Sesame*, the Aladdin matmûrah of relative logic had been nearly opened to my mind's eye... (Peirce, 1933b, p. 510)

Peirce struggled to overcome the moribund traditional logic both by studying the true medieval or scholastic logic and by developing his logic of relations and his quantificational logic.

### 7.4.3   Peirce as a teacher

During the years 1879–1884, Peirce had a part-time lectureship in Logic at Johns Hopkins. He had written one major paper in logic in 1870 and some philosophical papers on applied scientific method in 1877–78. The years at Johns Hopkins were fruitful for Peirce; he wrote four of his five most important papers in logic then. (See Appendix A for an annotated list of his main papers in logic.) In his courses on logic, which included both what we would call mathematical logic and what we might call philosophy of science, he used his own papers as texts, and he tried to bring students to new problems of inquiry at every stage of his courses.

The few memoirs of Peirce as a teacher, written by his students, lead us to think of long periods of questioning and inquiry interspersed with shorter periods of illuminating insight (Jastrow, 1916; Ladd-Franklin, 1916). The descriptions given in these memoirs should suggest to us the discovery method or even the 'Moore method'. Indeed, Peirce himself said that in teaching mathematics, "The Socratic method is best, but it is very difficult to get that method to work right." And it does seem that Benjamin Peirce was very Socratic when he tutored young Charles. Of course, as is well-known in regard to the Moore method, many students are habituated to the memorization of rules, and such students complain about the Socratic method of instruction. Jastrow's memoir, quoted above in Section 6.2, precedes an anecdote about Peirce and Jastrow, and both are provided here:

> The young men in my group who were admitted to his circle found him a most agreeable companion. The terms of equality upon which he met us were not in the way of flattery, for they were too spontaneous and sincere. We were members of his "scientific" fraternity; greetings were brief, and we proceeded to the business that brought us together, in which he and we found more pleasure than in anything else. This type of cooperation and delegation of responsibility came as near to a pedagogical device as any method that he used. One instance of it stands out with embarrassing clearness. To my consternation I was informed by Mr. Peirce that he would be absent at the time of the next lecture in logic, and that he would like me to present the next stage in the development of his topic to the class of graduate students. About half the hour was over, when Mr. Peirce walked in, took his place and insisted upon my concluding the exercise. I know of no more enlightening comment upon the atmosphere of the place and the day than that the procedure was accepted naturally by all concerned except myself. (Jastrow, 1916, p. 725)

Peirce had only a handful of students in most of his courses; John Dewey dropped Peirce's advanced logic course because it was too mathematical (Houser, 1986, p. lxi).

Had Peirce been able to continue at Johns Hopkins, he probably would have founded a school of logic—an active group but not an institution. We may take the 1883 volume of *Studies in Logic by Members of the Johns Hopkins University* as indicative of the kind of work that would have been produced by the group. Conceivably, a great deal of logic would have been developed earlier than it was and in somewhat different ways. As Jastrow wrote in his memoir of Peirce,

> The "Algebra of Logic" was an expert tool usable only by the expert and extending the scope of the logical grasp. Deeply mathematical, his thinking had not the trace of a scholastic quality; there was no love of the tool for its own sake, but an admiration of its cutting edge as the issue of human care and skill. ... His command of the history of science was encyclopedic in the best sense of the word. The hypotheses of the great thinkers of the past were transformed into logical exercises for the present-day student. ... The irrelevant was discarded, the significant composition revealed. The chips fell away and the statue in the block appeared. This sense of masterly analysis

accomplished with neatness and dispatch,—all seemingly easy, but actually the quality of the highest type of keen thinking—remains as the central impression of a lecture by Professor Peirce. (Jastrow, 1916, p. 723)

### 7.4.4  Peirce's mathematics textbooks

During the 1890s, Peirce wrote and tried to publish a set of school mathematics textbooks. He produced manuscripts for arithmetic, geometry, and algebra (Peirce, 1976a; 1976b). These books are pregnant with implications for how Benjamin Peirce must have tutored Charles. Imagine that a new mathematics (or perhaps "new math") textbook were to be written by a research mathematician for a particularly highly gifted child.[6] There are different number bases in the arithmetic, deep topological ideas in the geometry, etc. Nevertheless, many of the examples in these manuscripts can be acted out or demonstrated with simple material objects. There is no evidence that Peirce actually tried this material on children, although that is certainly possible. The companies either refused his manuscripts outright or else demanded such extensive changes that Peirce could not make them in his own good conscience. One can only speculate on how much mathematics education would have been affected if Peirce's textbooks had been published and used. It might be worthwhile to look at these books, to see whether there is material of value for contemporary mathematics education.

## 7.5  Concluding remarks

Peirce was a misfit in the nineteenth century. Had he lived fifty years later, he might have invented some of the important reform movements in twentieth century mathematics education. He would certainly have been very interested in the computer science developments described in Section 7.2 above; he would also have been very interested in the use of computers as an aid to developing theoric, diagrammatic reasoning in students—especially in LOGO and its successors.

To recapitulate the argument of this book: In Chapter 1, the purpose, the subject, and the method of the inquiry are introduced. In Chapters 2 and 3, a two stage method is applied in detail to some of Peirce's most important work in mathematical logic; this technical establishment of the mathematical content in the selections from Peirce provides a foundation for the rest of the book. In Chapters 4 and 5, Peirce's ideas on what we might

---

[6]Of course, one of the failings of the so-called "new math" of the mid-twentieth century was that the research mathematicians who wanted to share their great interest and enjoyment in mathematics were not available to do the teaching, which had to be left to teachers who were not familiar with the deeper aspects of the ideas involved.

call today the philosophy of mathematics and the philosophy of science are discussed; these ideas are relevant because Peirce was strongly motivated by his philosophical ideas to pursue his work in logic. In Chapter 6, Peirce's place in the history of mathematical logic and the lack of recognition for his work are discussed; the development of this discussion involves all of the preceding material, but it rests ultimately on the technical material of Chapters 2 and 3. In this final Chapter 7, some further connections from Peirce's life and ideas to present-day ones are briefly indicated.

Charles Sanders Peirce was a philosophical genius, a mathematician of very great talent, and a professional scientist with the Coast and Geodetic Survey for much of his working life. His philosophical development was the result of his own original work; his mathematical development was the result of his tutelage by and later collaboration with his mathematician father; and he achieved an eminent reputation as an applied scientific thinker in astronomy and geodesy. Peirce's development in these three different ways of the intellect—speculative philosophy, pure mathematics, and applied science—made him uniquely qualified to bring all of them to bear on a discipline he called "logic". He never held a full-time position in a university faculty. He tried without much success to make his living in later life as an author. Peirce was one of those rare individuals whose life exemplifies the curriculum given by Plato in the seventh book of the *Republic*: that is to say, the most difficult and concentrated mathematical and scientific studies are the necessary training for a philosopher (Plato, 1968, pp. 201–211).

In this book we have tried to follow a scientific method in the history of logic, a method we hope has been in the spirit of Peirce, van der Waerden, and Weil. Claims about ideas and influence have been supported with material from original sources; speculations, when offered, have been clearly indicated as such. Ultimately, we have tried to honor the sentiment expressed by Leibniz, as quoted at the beginning of Chapter 1 and in Section 6.1.3 of the dialogue in Chapter 6:

> Its use is not just that History may give everyone his due and that others may look forward to similar praise, but also that the art of discovery be promoted and its method known through illustrious examples. (Weil, 1978, p. 227)

# Appendix A

# Main Papers on Logic and Mathematics

The numbers are the authoritative bibliographic numbers for Peirce's publications, which are given in *A Comprehensive Bibliography of the Published Works of Charles Sanders Peirce* (Ketner, 1986).

P 52: Description of a Notation for the Logic of Relatives, Resulting from an Amplification of the Conceptions of Boole's Calculus of Logic. *Memoirs of the American Academy of Arts and Sciences*, n. s. 9: 317–378, 1870.

In *Collected Papers of Charles Sanders Peirce, Volume III, Exact Logic (Published Papers)*, ed. C. Hartshorne and P. Weiss, pp. 27–98. Cambridge, Harvard University Press, 1933a.

In *Writings of Charles S. Peirce: A Chronological Edition, Volume 2, 1867–1871*, ed. E.C. Moore et al., pp. 359–429. Bloomington, Indiana University Press, 1984.

P 52, the paper above, is the first of Peirce's five major papers in mathematical logic. In it, Peirce developed his algebra of relations and classes. The paper is discussed in Chapter 2, Sections 2.3–2.7 of this book.

P 90: On the Application of Logical Analysis to Multiple Algebra. *Proceedings of the American Academy of Arts and Sciences*, 10: 392–394, 1875.

In *Collected Papers of Charles Sanders Peirce, Volume III, Exact Logic (Published Papers)*, ed. C. Hartshorne and P. Weiss, pp. 99–101. Cambridge, Harvard University Press, 1933a.

In *Writings of Charles S. Peirce: A Chronological Edition, Volume 3, 1872–1878*, ed. C.J.W. Kloesel et al., pp. 177–179. Bloomington, Indiana University Press, 1986a.

P 90, the brief paper above, contains the first statement of Peirce's representation theorem, which he treated more thoroughly in item [P 188c] below.

P 167: On the Algebra of Logic. *American Journal of Mathematics*, 3: 15–57, 1880.

In *Collected Papers of Charles Sanders Peirce, Volume III, Exact Logic (Published Papers)*, ed. C. Hartshorne and P. Weiss, pp. 104–157. Cambridge, Harvard University Press, 1933a.

In *Writings of Charles S. Peirce: A Chronological Edition, Volume 4, 1879–1884*, ed. C.J.W. Kloesel et al., pp. 163–209. Bloomington, Indiana University Press, 1986b.

P 167, the paper above, is the second of Peirce's five major papers in mathematical logic. It contains Peirce's main contribution to lattice theory and an unsuccessful attempt to deal with problems requiring quantification.

P 188c: On the Relative Forms of the Algebras. *American Journal of Mathematics*, 4: 221–225, 1881.

In *Collected Papers of Charles Sanders Peirce, Volume III, Exact Logic (Published Papers)*, ed. C. Hartshorne and P. Weiss, pp. 171–175. Cambridge, Harvard University Press, 1933a.

In *Writings of Charles S. Peirce: A Chronological Edition, Volume 4, 1879–1884*, ed. C.J.W. Kloesel et al., pp. 319–322. Bloomington, Indiana University Press, 1986b.

P 188c, the paper above, is one of the endnotes to the reprinting of Benjamin Peirce's 1870 *Linear Associative Algebra* (B. Peirce, 1881), which Charles S. Peirce edited. It contains Peirce's proof of his representation theorem: Every associative algebra of dimension n is isomorphic to a subalgebra of $P_{n+1}$. The proof is analyzed in Section 3.2 of this book.

P 220: *Brief Description of the Algebra of Relatives*. Baltimore, privately printed, 1882.

In *Collected Papers of Charles Sanders Peirce, Volume III, Exact Logic (Published Papers)*, ed. C. Hartshorne and P. Weiss, pp. 180–186. Cambridge, Harvard University Press, 1933a.

In *Writings of Charles S. Peirce: A Chronological Edition, Volume 4, 1879–1884*, ed. C.J.W. Kloesel et al., pp. 328–333. Bloomington, Indiana University Press, 1986b.

P 220, the short paper above, is the third of Peirce's five major papers in mathematical logic. This paper of 1882 contains Peirce's matrix representation of his relation algebra, but the scalars are not truth values, being taken as real or complex numbers. It is discussed in Section 3.2 of this book.

P 268d: Note B: The Logic of Relatives. In *Studies in Logic, By*

*Members of the Johns Hopkins University*, ed. C.S. Peirce, pp. 187–203. Boston, Little Brown, 1883. Reprinted: Amsterdam, John Benjamins, 1983.

In *Collected Papers of Charles Sanders Peirce, Volume III, Exact Logic (Published Papers)*, ed. C. Hartshorne and P. Weiss, pp. 195–209. Cambridge, Harvard University Press, 1933a.

In *Writings of Charles S. Peirce: A Chronological Edition, Volume 4, 1879–1884*, ed. C.J.W. Kloesel et al., pp. 453–466. Bloomington, Indiana University Press, 1986b.

P 268d the paper above, is the fourth of Peirce's five major papers in mathematical logic. In it, Peirce gave the matrix interpretation of his algebra of relatives. He used $\sum$ and $\prod$ to stand for sums and products of truth values with respect to atomic formulas and used variables for individual elements of the universe, as discussed in Section 3.3 of this book.

P 296: On the Algebra of Logic: A Contribution to the Philosophy of Notation. *American Journal of Mathematics*, 7: 180–202, 1885.

In *Collected Papers of Charles Sanders Peirce, Volume III, Exact Logic (Published Papers)*, ed. C. Hartshorne and P. Weiss, pp. 210–238. Cambridge, Harvard University Press, 1933a.

P 296 the paper above, is the last and greatest of Peirce's five major papers in mathematical logic. It contains the main elements of a framework for first- and second-order logic, as discussed in Section 3.5 of this book. The shift is made from the quantifiers $\sum$ and $\prod$ conceived as actually being summation and production of truth values to quantification conceived as only being like summation and production. "Quantifier" is used to mean the entire string of quantifiers in prenex form.

P 620: The Regenerated Logic. *The Monist*, 7: 19–40, 1896.

In *Collected Papers of Charles Sanders Peirce, Volume III, Exact Logic (Published Papers)*, ed. C. Hartshorne and P. Weiss, pp. 266–287. Cambridge, Harvard University Press, 1933a.

P 620, the paper above, is a review of Ernst Schröder's work on "exact logic", Peirce gave an explanation of his views on the relations between mathematics, logic, and philosophy, as discussed in Sections 1.4 and 4.2 of this book. "Quantifer" is used in the present-day sense of a single symbol for 'all' or 'some' attached to a variable for an individual. This paper also contains the clearest statement of Peirce's notion of logical necessity, which is his informal version of Tarski's twentieth century rigorous concept of logical consequence, as discussed in Section 3.6.

# Appendix B

# Chronology of Peirce's Pragmatism

ca 150 BC—The Greek historian Polybius writes *pragmatike historia* to be instructive and useful to the living.

ca 0—A Roman *pragmaticus* is one skilled in the law who furnishes orators and advocates with the principles on which they base their speeches.

ca 1300—Duns Scotus proposes a modified realism, a theory of true universals inherent in particulars.

ca 1790—Kant uses "practical" and "pragmatic"; according to Peirce, *practisch*—"belonging in a region of thought where no mind of the experimentalist type can ever make sure of solid ground under his feet", and *pragmatisch*— "expressing relation to some definite human purpose". (Peirce, 1934, p. 274)

1839—Peirce is born.

1861—Peirce becomes a research scientist at the Coast Survey.

1870—Peirce publishes his major paper on the logic of relations.

1871-75—Peirce, William James, Chauncey Wright, Oliver Wendell Holmes, et al. discuss pragmatic ideas in the Cambridge 'Metaphysical Club'; lawyer Nicholas St. John Green stresses Alexander Bain's definition of belief as "that upon which a man is prepared to act."

1872—In November, Peirce states the pragmatic maxim in a talk at a Metaphysical Club meeting, using the word "pragmatism".

1877-78—Peirce publishes the material of the 1872 talk as 'The Fixation of Belief' and 'How to Make Our Ideas Clear', two articles in *Popular Science Monthly*. The pragmatic maxim is given in the second article, but the word "pragmatic" is not used there.

1879—Peirce accepts a position as a part-time lecturer on logic at Johns Hopkins.

1881—Peirce publishes his representation theorem.

1883—Peirce publishes his discovery of the quantifiers.

1884—Peirce is removed from his position at Johns Hopkins.

1885—Peirce publishes his great paper on first- and second-order logic.

1891—Peirce is forced to resign from the Coast and Geodetic Survey.

1891–93—Peirce publishes a series of five articles on philosophy in *The Monist*.

1898—Peirce gives the Cambridge Conference lectures.

1898—William James credits Peirce's 1878 article, 'How to Make Our Ideas Clear', in his talk, 'Philosophical Conceptions and Practical Results'. James paraphrases the maxim while implying that he is quoting it, substituting "effects" (material consequences) for "conceivable effects" (bearing upon conduct).

1903—Peirce gives the Lowell Lectures on pragmatism.

1905–06—Peirce responds to James and others with three articles ('What Pragmatism Is', 'Issues of Pragmaticism', and 'Prologomena to an Apology for Pragmaticism') in *The Monist*.

1908—Arthur O. Lovejoy discriminates 'The Thirteen Pragmatisms'.

1914—Peirce dies.

1916—The *Journal of Philosophy* publishes a memorial issue devoted to Peirce, including a bibliography of his publications, compiled by Morris Cohen, and an article, 'The Pragmatism of Peirce', by John Dewey.

1923—Cohen edits *Chance, Love, and Logic*, which includes the 1877–78 *Popular Science Monthly* series, the 1891–93 *Monist* series, and Dewey's 1916 essay.

1931–35—Hartshorne and Weiss edit volumes 1 through 6 of *Collected Papers*.

1958—Burks edits volumes 7 and 8 of *Collected Papers*.

1976—Eisele edits Peirce's mathematics textbook manuscripts and other mathematical material as, *The New Elements of Mathematics*.

1982–present—The Peirce Edition Project begins publishing a chronological edition. Thirty volumes are planned; vol. 1 (1857–1866), vol. 2 (1867–1871), vol. 3 (1872–1878), vol. 4 (1879–1884), and vol. 5 (1884–1886) are out by 1992.

1985—Eisele edits Peirce's history of science manuscripts as, *Historical Perspectives on Peirce's Logic of Science: A History of Science*.

1989—The Peirce Sesquicentennial International Congress is held at Harvard.

# Appendix C

# Cited Literature

Beth, E.: *The Foundations of Mathematics, Second Revised Edition.* Amsterdam, North-Holland Publishing, 1968.

Bochenski, I.M.: *A History of Formal Logic.* New York, Chelsea, 1970.

Boole, G.: *A Mathematical Analysis of Logic: Being an Essay Towards a Calculus of Deductive Reasoning.* Cambridge, Macmillan, 1847. Reprinted: Oxford, England, Basil Blackwell, 1965.

Boole, G.: *An Investigation of the Laws of Thought.* Cambridge, Macmillan, 1854. Reprinted: New York, Dover, 1958.

Boswell, J.: *Life of Samuel Johnson LL.D.* In *Great Books of the Western World, Volume 44.* Chicago, Encyclopaedia Brittanica, 1952.

Bourbaki, N.: *Éléments d'histoire des mathématiques.* Paris, Hermann, 1960.

Bowen, F.: *The Metaphysics of Sir William Hamilton.* Boston, John Allyn, 1873.

Brady, G.: *Peirce's Introduction of the Quantifiers.* M.A. Thesis, University of Chicago, 1990.

Burks, A.W.: Logic, Biology and Automata—Some Historical Reflections. *International Journal of Man-Machine Studies,* 7: 297–312, 1975.

Carnap, R.: Intellectual Autobiography. In *The Philosophy of Rudolf Carnap,* ed. P.A. Schilpp, pp. 3–84. La Salle, Illinois, Open Court, 1963.

Church, A.: *Introduction to Mathematical Logic, Volume I.* Princeton, Princeton University Press, 1956.

Codd, E.F.: *Propagation, Computation, and Construction in Two-Dimensional Cellular Spaces.* PhD. Thesis, University of Michigan, 1965.

Codd, E.F.: A Relational Model of Data for Large Shared Data Banks. *Communications of the ACM,* 13: 377–387, 1970.

Codd, E.F.: Relational Database: A Practical Foundation for Productivity. In *ACM Turing Award Lectures: The First Twenty Years,* ed. R.L. Ashenhurst, pp. 391–409. New York, Addison-Wesley / ACM Press, 1987.

Codd, E.F.: The relational model and beyond. *Computer Language*, March 1988: 89–97.

Codd, E.F.: Relational Philosopher. *DBMS*, December 1990: 34–40, 60.

Cohen, J.: A View of the Origins and Development of Prolog. *Communications of the ACM*, 31: 26–36, 1988.

Cohen, M.R.: Preface. In *Chance, Love, and Logic*, ed. M.R. Cohen, pp. iii–iv. New York, Harcourt, Brace, 1923. Reprinted: New York, George Braziller, 1956.

Curry, H.B.: *Foundations of Mathematical Logic*. New York, McGraw-Hill, 1963. Reprinted: New York, Dover, 1977.

Dante: *The Divine Comedy: Inferno*, trans. C.S. Singleton. Princeton, New Jersey, Princeton University Press, 1980.

De Morgan, A.: *On the Syllogism and Other Logical Writings*, ed. P. Heath. London, Routledge and Kegan Paul, 1966.

Dedekind, R.: *Essays on the Theory of Numbers*, trans. W.W. Beman. New York, Dover, 1963.

Dewey, J.: The Pragmatism of Peirce. In *Chance, Love, and Logic*, ed. M.R. Cohen, pp. 301–308. New York, Harcourt, Brace, 1923. Reprinted: New York, George Braziller, 1956.

Dewey, J.: The Development of American Pragmatism. In *Philosophy and Civilization*, pp. 13–35. New York, Capricorn Books, 1963.

Dostoevsky, F.: *The Brothers Karamazov*, trans. R. Pevear and L. Volokhonsky. San Francisco, North Point Press, 1990.

Drake, K.C. and Hess, P.: Abduction: A Numeric Kowledge Acquisition Approach. *PC AI*, September / October 1990: 58–61.

Dreben, B. and van Heijenoort, J.: Introductory note to *1929, 1930* and *1930a*. In *Kurt Gödel: Collected Works, Volume I, Publications 1929–1936*, ed. S. Feferman et al., pp. 44–59. New York, Oxford University Press, 1986.

Dummett, M.: Review of Boole, *Studies in logic and probability*. *Journal of Symbolic Logic*, 24: 203–209, 1959.

Ewald, W.B.: The Logic of the Thing. *Times Literary Supplement*, 8 June 1990: 599–600.

Fenstad, J.E.: Thoralf Albert Skolem in Memoriam. In *Selected Works in Logic*, ed. J.E. Fenstad, pp. 9–15. Oslo, Universitetsforlaget, 1970.

Fisch, M.H.: Introduction. In *Writings of Charles S. Peirce: A Chronological Edition, Volume 1, 1857–1866*, ed. M. Fisch et al., pp. xv–xxxv. Bloomington, Indiana University Press, 1982.

Fisch, M.H.: The Decisive Year and Its Early Consequences. In *Writings of Charles S. Peirce: A Chronological Edition, Volume 2, 1867–1871*, ed. E.C. Moore et al., pp. xxi–xxxvi. Bloomington, Indiana University Press, 1984.

Fisch, M.H.: Introduction. In *Writings of Charles S. Peirce: A Chronological Edition, Volume 3, 1872–1878*, ed. C.J.W. Kloesel et al., pp. xxi–xxxvii. Bloomington, Indiana University Press, 1986a.

Fisch, M.H.: *Peirce, Semeiotic, and Pragmatism*, ed. K.L. Ketner and C.J.W. Kloesel. Bloomington, Indiana University Press, 1986b.

Fisch, M.H. and Kloesel, C.J.W.: Peirce and the Florentine Pragmatists: His Letter to Calderoni and a New Edition of His Writings. *Topoi*, 1: 68–73, 1982.

Frege, G.: *Begriffsschrift*, a formula language, modeled upon that of arithmetic, for pure thought. In *From Frege to Gödel: A Sourcebook in Mathematical Logic, 1879–1931*, ed. J. van Heijenoort, pp. 1–82. Cambridge, Harvard University Press, 1977.

Gardner, M.: *Logic Machines and Diagrams*. Chicago, University of Chicago Press, 1982.

Gibbon, E.: *The Decline and Fall of the Roman Empire, Volume III*, ed. J.B. Bury. London, Metheun and Company, 1909. Reprinted: New York, AMS Press, 1974.

Gödel, K.: *Kurt Gödel: Collected Works, Volume I, Publications 1929–1936*, ed. S. Feferman et al. New York, Oxford University Press, 1986.

Goldfarb, W.: Introduction. In *Jacques Herbrand: Logical Writings*, ed. W. Goldfarb, pp. 1–20. Cambridge, Harvard University Press, 1971.

Goldfarb, W.: Logic in the Twenties: The Nature of the Quantifier. *Journal of Symbolic Logic*, 44: 351–368, 1979.

Hamilton, W.: *Discussions on Philosophy and Literature, Education and University Reform*. New York, Harper and Brothers, 1853.

Hardwick, C.S.: Introduction. In *Semiotic and Significs: The Correspondence between Charles S. Peirce and Lady Welby*, ed. C.S. Hardwick, pp. xv–xxxiv. Bloomington, Indiana University Press, 1977.

Harris, W.H. and Levey, J.S.: *The New Columbia Encyclopedia*. New York, Columbia University Press, 1975.

Heath, P.: Editor's Introduction. In *On the Syllogism and Other Logical Writings*, ed. P. Heath, pp. vii–xxxi. London, Routledge and Kegan Paul, 1966.

Herbrand, J.: *Jacques Herbrand: Logical Writings*, ed. W. Goldfarb. Cambridge, Harvard University Press, 1971.

Hilbert, D.: *Foundations of Geometry, Second Edition*. Lasalle, Illinois, Open Court, 1971.

Hilbert, D. and Ackermann, W.: *Principles of Mathematical Logic*, ed. and trans. R.E. Luce. New York, Chelsea Publishing, 1950.

Hoffstadter, D.: *Gödel, Escher, Bach: an Eternal Golden Braid*. New York, Basic Books, 1979.

Houser, N.: Introduction. In *Writings of Charles S. Peirce: A Chronological Edition, Volume 4, 1879–1884*, ed. C.J.W. Kloesel et al., pp. xix–lxx. Bloomington, Indiana University Press, 1986.

James, W.: *The Writings of William James*, ed. J.J. McDermott. New York, Modern Library, 1968.

Jastrow, J.: Charles S. Peirce as a Teacher. *The Journal of Philosophy, Psychology, and Scientific Methods*, 13: 723–726, 1916.

Kant, I.: *Immanuel Kant's Critique of Pure Reason*, ed. and trans. N.K. Smith. New York, St. Martin's Press, 1958.

Kennedy, H.C.: 'Biographical sketch of Giuseppe Peano' and [Introductions to individual papers]. In *Selected works of Giuseppe Peano*, ed. and trans. H.C. Kennedy, pp. 3–10, etc. Toronto, University of Toronto Press, 1973.

Ketner, K.L.: *A Comprehensive Bibliography of the Published Works of Charles Sanders Peirce with a Bibliography of Secondary Studies, Second edition, revised*. Bowling Green, Philosophy Documentation Center, 1986.

Ketner, K.L.: Charles Sanders Peirce: Introduction. In *Classical American Philosophy; Essential Readings and Interpretive Essays*, ed. J.J. Stuhr, pp. 13–25. New York, Oxford University Press, 1987.

Kneale, W. and Kneale, M.: *The Development of Logic*. London, Oxford University Press, 1962.

Kowalski, R.A.: The Early Years of Logic Programming. *Communications of the ACM*, 31: 38–43, 1988.

Ladd, C.: On the Algebra of Logic. In *Studies in Logic by Members of the Johns Hopkins University*, ed. C.S. Peirce, pp. 17–71. Boston, Little Brown, 1883. Reprinted: Amsterdam, John Benjamins, 1983.

Ladd-Franklin, C.: Charles S. Peirce at the Johns Hopkins. *The Journal of Philosophy, Psychology, and Scientific Methods*, 13: 715–722, 1916.

Laita, L.M.: Boolean Algebra and its Extra logical Sources: The Testimony of Mary Everest Boole. *History and Philosophy of Logic*, 1: 37–60, 1980.

Lovejoy, A.O.: *The Thirteen Pragmatisms and Other Essays*. Baltimore, Johns Hopkins Press, 1963.

Löwenheim, L.: Über Möglichkeiten im Relativkalkül. *Mathematische Annalen*, 76: 447–470, 1915.

Löwenheim, L.: On possibilities in the calculus of relatives. Translation of (Löwenheim, 1915), in *From Frege to Gödel: A Sourcebook in Mathematical Logic, 1879–1931*, trans. S. Bauer-Mengelberg, ed. J. van Heijenoort, pp. 232–251. Cambridge, Harvard University Press, 1977.

MacHale, D.: *George Boole: His Life and Work*. Dublin, Boole Press, 1985.

McKeon, R.P.: The Battle of the Books. In *The Knowledge Most Worth Having*, ed. W.C. Booth, pp. 173–202. Chicago, University of Chicago Press, 1967.

Mandelbaum, A.: A Glossary. In *The Aeneid of Virgil*, trans. A. Mandelbaum, pp. 351–407. University of California Press, Berkeley, 1981.

Marquand, A.: A Machine for Producing Syllogistic Variations. In

*Studies in Logic by Members of the Johns Hopkins University*, ed. C.S. Peirce, pp. 13–16. Boston, Little Brown, 1883. Reprinted: Amsterdam, John Benjamins, 1983.

Merrill, D.D.: De Morgan, Peirce, and the Logic of Relations. *Transactions of the Charles S. Peirce Society*, 14: 247–284, 1978.

Merrill, D.D.: The 1870 Logic of Relatives Memoir. In *Writings of Charles S. Peirce: A Chronological Edition, Volume 2, 1867–1871*, ed. E.C. Moore et al., pp. xlii–xlviii. Bloomington, Indiana University Press, 1984.

Mitchell, O.H.: On a New Algebra of Logic. In *Studies in Logic by Members of the Johns Hopkins University*, ed. C.S. Peirce, pp. 72–106. Boston, Little Brown, 1883. Reprinted: Amsterdam, John Benjamins, 1983.

Moore, G.H.: From Frege to Gödel: A Source Book in Mathematical Logic, 1879–1931. *Historia Mathematica*, 4: 468–471, 1977.

Moore, G.H.: The Emergence of First-Order Logic. In *History and Philosophy of Modern Mathematics*, ed. W. Asprey and P. Kitcher, pp. 95–135. Minneapolis, University of Minnesota Press, 1988.

Nagel, E. and Newman, J.R.: *Gödel's Proof*. New York, New York University Press, 1958.

Naimark, M.A. and Stern, A.I.: *Theory of Group Representations*. Berlin, Springer-Verlag, 1982.

O'Rorke, P.: Review of AAAI 1990 Spring Symposium on AUTOMATED ABDUCTION. *ACM SIGART Bulletin*, 1: 12–17, 1990.

Palmer, G.H.: *The Autobiography of a Philosopher*. Boston, Houghton Mifflin, 1930.

Peano, G.: *Selected works of Giuseppe Peano*, ed. and trans. H.C. Kennedy. Toronto, University of Toronto Press, 1973.

Peirce, B.: *Linear Associative Algebra*. Washington City, privately printed, 1870.

Peirce, B.: Linear Associative Algebra. With Notes and Addenda by C. S. Peirce, son of the Author. *American Journal of Mathematics*, 4: 97–229, 1881.

Peirce, C.S. : Logical Machines. *American Journal of Psychology*, 1: 165–170, 1887.

Peirce, C.S. : Proximate. In *Dictionary of Philosophy and Psychology, vol. 2*, ed. J.M. Baldwin, pp. 373–374. New York, Macmillan, 1902.

Peirce, C.S.: *Collected Papers of Charles Sanders Peirce, Volume I, Principles of Philosophy*, ed. C. Hartshorne and P. Weiss. Cambridge, Harvard University Press, 1931.

Peirce, C.S.: *Collected Papers of Charles Sanders Peirce, Volume II, Elements of Logic*, ed. C. Hartshorne and P. Weiss. Cambridge, Harvard University Press, 1932.

Peirce, C.S.: *Collected Papers of Charles Sanders Peirce, Volume III, Exact Logic (Published Papers)*, ed. C. Hartshorne and P. Weiss.

Cambridge, Harvard University Press, 1933a.

Peirce, C.S.: *Collected Papers of Charles Sanders Peirce, Volume IV, The Simplest Mathematics*, ed. C. Hartshorne and P. Weiss. Cambridge, Harvard University Press, 1933b.

Peirce, C.S.: *Collected Papers of Charles Sanders Peirce, Volume V, Pragmatism and Pragmaticism*, ed. C. Hartshorne and P. Weiss. Cambridge, Harvard University Press, 1934.

Peirce, C.S.: *Collected Papers of Charles Sanders Peirce, Volume VI, Scientific Metaphysics*, ed. C. Hartshorne and P. Weiss. Cambridge, Harvard University Press, 1935.

Peirce, C.S.: *Collected Papers of Charles Sanders Peirce, Volume VII, Science and Philosophy*, ed. A. Burks. Cambridge, Harvard University Press, 1958a.

Peirce, C.S.: *Collected Papers of Charles Sanders Peirce, Volume VIII, Reviews, Correspondence, and Bibliography*, ed. A. Burks. Cambridge, Harvard University Press, 1958b.

Peirce, C.S.: *The New Elements of Mathematics, Volume I, Arithmetic*, ed. C. Eisele. The Hague, Mouton Publishers, 1976a.

Peirce, C.S.: *The New Elements of Mathematics, Volume II, Algebra and Geometry*, ed. C. Eisele. The Hague, Mouton Publishers, 1976b.

Peirce, C.S.: *The New Elements of Mathematics, Volume III, Mathematical Miscellanea*, ed. C. Eisele. The Hague, Mouton Publishers, 1976c.

Peirce, C.S.: *The New Elements of Mathematics, Volume IV, Mathematical Philosophy*, ed. C. Eisele. The Hague, Mouton Publishers, 1976d.

Peirce, C.S.: *Writings of Charles S. Peirce: A Chronological Edition, Volume 1, 1857–1866*, ed. M.H. Fisch et al. Bloomington, Indiana University Press, 1982.

Peirce, C.S.: *Writings of Charles S. Peirce: A Chronological Edition, Volume 2, 1867–1871*, ed. E.C. Moore et al. Bloomington, Indiana University Press, 1984.

Peirce, C.S.: *Historical Persectives on Peirce's Logic of Science*, ed. C. Eisele. Berlin, Mouton, 1985.

Peirce, C.S.: *Writings of Charles S. Peirce: A Chronological Edition, Volume 3, 1872–1878*, ed. C.J.W. Kloesel et al. Bloomington, Indiana University Press, 1986a.

Peirce, C.S.: *Writings of Charles S. Peirce: A Chronological Edition, Volume 4, 1879–1884*, ed. C.J.W. Kloesel et al. Bloomington, Indiana University Press, 1986b.

Peirce, C.S. and Welby, V.: *Semiotic and Significs: The Correspondence between Charles S. Peirce and Lady Welby*, ed. C.S. Hardwick. Bloomington, Indiana University Press, 1977.

Peng, Y. and Reggia, J.: *Abductive Inference Models for Diagnostic Problem Solving*. New York, Springer-Verlag, 1990.

Plato: *The Republic of Plato*, trans. A. Bloom. New York, Basic Books, 1968.

Plato: *The Trial and Death of Socrates*, trans. G.M.A. Grube. Indianapolis, Indiana, Hackett Publishing Company, 1975.

Plutarch: Numa. In *Plutarch's Lives, Volume I*, trans. B. Perrin, pp. 305–401. Cambridge, Harvard University Press, 1914. Reprinted: 1982.

Plutarch: Alexander. In *Plutarch's Lives, Volume VII*, trans. B. Perrin, pp. 223–439. Cambridge, Harvard University Press, 1919. Reprinted: 1986.

Putnam, H.: Peirce the Logician. *Historia Mathematica*, 9: 290–301, 1982.

Quine, W.: Collected Papers of Charles Sanders Peirce.—Volume III: Exact Logic. *Isis*, 22: 285–297, 1934a.

Quine, W.: Collected Papers of Charles Sanders Peirce.—Volume IV: The Simplest Mathematics. *Isis*, 22: 551–553, 1934b.

Quine, W.: Whitehead and the Rise of Modern Logic. In *The Philosophy of Alfred North Whitehead*, ed. P.A. Schilpp, pp. 125–163. New York, Tudor Publishing Company, 1951.

Quine, W.: Preface. In Clark, J.T.: *Conventional Logic and Modern Logic*, pp. V–VII. Woodstock, Maryland, Woodstock College Press, 1952.

Quine, W.: In the Logical Vestibule. *Times Literary Supplement*, 12 July 1985: 767.

Roberts, D.D.: *The Existential Graphs of Charles S. Peirce*. The Hague, Mouton, 1973.

Robin, R.S.: *Annotated Catalogue of the Papers of Charles S. Peirce*. Amherst, University of Massachusetts Press, 1967.

Robin, R.S.: The Peirce Papers: A Supplementary Catalogue. *Transactions of the Charles S. Peirce Society*, 7: 37–57, 1971.

Robinson, J.A.: Theorem-Proving on the Computer. *Journal of the Association for Computing Machinery*, 10: 163–174, 1963.

Robinson, J.A.: A Machine-Oriented Logic Based on the Resolution Principle. *Journal of the Association for Computing Machinery*, 12: 23–41, 1965.

Robinson, J.A.: Logic and Logic Programming. *Communications of the ACM*, 35: 40–65, 1992.

Russell, B.: *Our Knowledge of the External World*. London, George Allen and Unwin, 1926.

Russell, B.: *Mysticism and Logic*. New York, Barnes and Noble, 1971.

Russell, B.: Foreword. In Feibleman, J.K.: *An Introduction to the Philosophy of Charles S. Peirce*, pp. xv–xvi. Cambridge, MIT Press, 1969.

Scholz, H.: *A Concise History of Logic*. New York, Philosophical Library, 1961.

Schröder, E.: Review of Frege, *Begriffsschrift*. In *Gottlob Frege: Conceptual Notation and Related Articles*, ed. T.W. Bynum, pp. 218–232. London, Oxford University Press, 1972.

Skolem, T.: Logisch-kombinatorische Untersuchungen über die Erfüllbarkeit oder Beweisbarkeit mathematischer Sätze nebst einem Theoreme über dichte Mengen. *Videnskapsselskapet skrifter, I. Matematisk-naturvidenskabelig klasse*, 4, 1920.

Skolem, T.: Einige Bemerkungen zur axiomatische Begründung der Mengenlehre. In *Matematikerkongressen i Helsingfors den 4–7 Juli 1922, Den femte skandinaviska matematikerkongressen, Redogörelse*, pp. 217–232. Helsinki, Akademiska Bokhandeln, 1923.

Skolem, T.: Logico-combinatorial investigations in the satisfiability or provability of mathematical propositions: A simplified proof of a theorem by L. Löwenheim and generalizations of the theorem. Translation of (Skolem, 1920), in *From Frege to Gödel: A Sourcebook in Mathematical Logic, 1879–1931*, trans. S. Bauer-Mengelberg, ed. J. van Heijenoort, pp. 254–263. Cambridge, Harvard University Press, 1977a.

Skolem, T.: Some remarks on axiomatized set theory. Translation of (Skolem, 1923), in *From Frege to Gödel: A Sourcebook in Mathematical Logic, 1879–1931*, trans. S. Bauer-Mengelberg, ed. J. van Heijenoort, pp. 291–301. Cambridge, Harvard University Press, 1977b.

Skolem, T.: On mathematical logic. In *From Frege to Gödel: A Sourcebook in Mathematical Logic, 1879–1931*, trans. S. Bauer-Mengelberg and D. Follesdal, ed. J. van Heijenoort, pp. 512–524. Cambridge, Harvard University Press, 1977c.

Sowa, J.F.: *Conceptual Structures*. Reading, Massachusetts, Addison-Wesley, 1984.

Swift, J.: The Battle of the Books. In *The Great Ideas Today 1971*, ed. R.M. Hutchins and M.J. Adler, pp. 380–401. Chicago, Encyclopaedia Brittanica, 1971.

Tarski, A.: On the Concept of Logical Consequence. In *Logic, Semantics, Metamathematics: Papers from 1923 to 1938, second edition*, trans. J.H. Woodger, ed. J. Corcoran, pp. 409–420. Indianapolis, Hackett, 1983.

Thagard, P.: Review of Abductive Inference Models for Diagnostic Problem Solving. *ACM SIGART Bulletin*, 2: 72–75, 1991.

Thayer, H.S.: *Meaning and Action: A Critical History of Pragmatism*. Indianapolis, Hackett, 1981a.

Thayer, H.S.: Pragmatism: A Reinterpretation of the Origins and Consequences. In *Pragmatism, Its Sources and Prospects*, ed. R.J. Mulvaney and P.M. Zeltner, pp. 1–20. Columbia, University of South Carolina Press, 1981b.

Thiel, C.: Leopold Löwenheim: Life, Work, and Early Influence. In *Logic Colloquium 76*, ed. R. Gandy and M. Hyland, pp. 235–252. Amsterdam, North-Holland Publishing, 1977.

van der Waerden, B.L.: *Science Awakening*. New York, Oxford University Press, 1961.

van der Waerden, B.L.: *A History of Algebra*. Berlin, Springer-Verlag,

1985.

van Heijenoort, J.: [Introductions to individual papers]. In *From Frege to Gödel: A Sourcebook in Mathematical Logic, 1879–1931*, ed. J. van Heijenoort. Cambridge, Harvard University Press, 1977a.

van Heijenoort, J.: Set-Theoretic Semantics. In *Logic Colloquium 76*, ed. R. Gandy and M. Hyland, pp. 183–190. Amsterdam, North-Holland Publishing, 1977b.

Virgil: *The Aeneid of Virgil*, trans. A. Mandelbaum. University of California Press, Berkeley, 1981.

Wang, H.: *From Mathematics to Philosophy*. London, Routledge and Kegan Paul, 1974.

Wang, H.: *Reflections on Kurt Gödel*. Cambridge, MIT Press, 1987.

Weil, A.: History of Mathematics: Why and How? *Proceedings of the International Congress of Mathematicians*. Helsinki, 1978.

Wiener, N.: *A comparison between the treatment of the algebra of relatives by Schroeder and that by Whitehead and Russell*. Ph.D. Thesis, Harvard University, 1913.

Wiener, N.: *Ex-Prodigy: My Childhood and Youth*. Cambridge, MIT Press, 1953.

Wiener, P.: *Evolution and the Founders of Pragmatism*. Cambridge, Harvard University Press, 1949. Reprinted: Philadelphia, University of Pennsylvania Press, 1972.

Wiener, P.: Pragmatism. In *Dictionary of the History of Ideas, Volume III*, ed. P. Wiener, pp. 551–570. New York, Charles Scribner's Sons, 1973.

Zeman, J.J.: *The Graphical Logic of C. S. Peirce*. Ph.D. Thesis, University of Chicago, 1964.

www.ingramcontent.com/pod-product-compliance
Lightning Source LLC
Chambersburg PA
CBHW020200090426

42734CB00008B/890